Springer Theses

Recognizing Outstanding Ph.D. Research

Aims and Scope

The series "Springer Theses" brings together a selection of the very best Ph.D. theses from around the world and across the physical sciences. Nominated and endorsed by two recognized specialists, each published volume has been selected for its scientific excellence and the high impact of its contents for the pertinent field of research. For greater accessibility to non-specialists, the published versions include an extended introduction, as well as a foreword by the student's supervisor explaining the special relevance of the work for the field. As a whole, the series will provide a valuable resource both for newcomers to the research fields described, and for other scientists seeking detailed background information on special questions. Finally, it provides an accredited documentation of the valuable contributions made by today's younger generation of scientists.

Theses are accepted into the series by invited nomination only and must fulfill all of the following criteria

- They must be written in good English.
- The topic should fall within the confines of Chemistry, Physics, Earth Sciences, Engineering and related interdisciplinary fields such as Materials, Nanoscience, Chemical Engineering, Complex Systems and Biophysics.
- The work reported in the thesis must represent a significant scientific advance.
- If the thesis includes previously published material, permission to reproduce this must be gained from the respective copyright holder.
- They must have been examined and passed during the 12 months prior to nomination.
- Each thesis should include a foreword by the supervisor outlining the significance of its content.
- The theses should have a clearly defined structure including an introduction accessible to scientists not expert in that particular field.

More information about this series at http://www.springer.com/series/8790

Tae Mok Gwon

A Polymer Cochlear Electrode Array: Atraumatic Deep Insertion, Tripolar Stimulation, and Long-Term Reliability

Doctoral Thesis accepted by
Seoul National University for the Degree of Doctor
of Philosophy in the Faculty of Electrical and Computer
Engineering, Seoul, Korea (Republic of)

 Springer

Author
Dr. Tae Mok Gwon
Department of Electrical and Computer
  Engineering
Seoul National University
Seoul
Korea (Republic of)

Supervisor
Prof. Sung June Kim
Department of Electrical and Computer
  Engineering
Seoul National University
Seoul
Korea (Republic of)

ISSN 2190-5053   ISSN 2190-5061   (electronic)
Springer Theses
ISBN 978-981-13-0471-2   ISBN 978-981-13-0472-9   (eBook)
https://doi.org/10.1007/978-981-13-0472-9

Library of Congress Control Number: 2018942163

Printed on acid-free paper

This Springer imprint is published by the registered company Springer Nature Singapore Pte Ltd.
part of Springer Nature
The registered company address is: 152 Beach Road, #21-01/04 Gateway East, Singapore 189721, Singapore

Supervisor's Foreword

Conventional cochlear implants have gained success through many scientific discoveries and engineering innovations and established their effect on impaired cochlear neural functions. Worldwide, there are tens of millions of people who need these implantable devices to hear sound and pursue basic life. And yet, in spite of technological advancements, current cochlear implants are too costly to expect all potential recipients to benefit. Therefore, a hope was there that some method will be available where economies of scale can apply and thus the most effective devices can be supplied at a cost low enough so all those in need can benefit. Current metal-based cochlear implants have limits in this aspect, and thus the search was on for a new type of cochlear implant using a new class of material and accompanying technology.

Polymers have been widely used in many biomedical applications due to their applicability to batch processing as well as their inherent material properties such as being flexible. Our group noted liquid crystal polymer (LCP), a class of thermoplastic polymer, as the one that can meet our needs for making low-cost neural prosthetic implantable devices. LCP's outstanding water absorption rate (<0.04%) is what makes it suitable for monolithic encapsulation of electronic components. As a polymer, the material allows easy passage of electromagnetic signals through it, which enables smaller, antennae-integrated implantable electronics package compared to that made of metals. LCPs are available in film form so that microelectronic fabrication techniques can be applied to neural interfaces that can be monolithically integrated with electronics on the same LCP substrate in a seamless way, thus enhancing long-term reliability as well as cost-effectiveness.

The work presented in this thesis is associated with the development of LCP-based cochlear electrode array for clinical use. The author has designed, fabricated, and evaluated LCP-based cochlear electrode array for an improved polymer-based cochlear implant. In this study, in contrast to a conventional cochlear implant which is composed of titanium-encased electronics and wire-based multichannel electrode array, the author aims to develop and evaluate an LCP-based cochlear implant using thin-film processes and microelectromechanical system (MEMS) technologies which are compatible with mass production. The thesis deals with three key topics: atraumatic deep insertion, tripolar stimulation, and long-term reliability. Atraumatic cochlear electrode array has become indispensable in state-of-the-art cochlear implants such as electric acoustic stimulation (EAS), wherein the preservation of residual hearing is significant. In this work, a novel tapered design of LCP-based cochlear electrode array is presented to fulfill such goals. Next, local tripolar stimulation using multilayered electrode sites are shown to achieve highly focused electrical stimulation to reduce channel interaction which should be avoided for high-density and pitch-recognizable cochlear implant. Lastly, this thesis addresses another vital issue of the long-term reliability of the polymer-based neural implants. After suggesting a new method of forming mechanical interlocking to improve polymer–metal adhesion, the author performs accelerating aging tests and analyzes comprehensive and systematic review to verify the method's efficacy.

These topics have been thoroughly examined through various in vitro and in vivo studies. Verification foretells the development of LCP-based cochlear electrode array for an atraumatic deep insertion, advanced stimulation, and long-term clinical implant. The new cochlear electrode arrays have been proven safe, effective, and reliable. The contents of this thesis contribute to expansion of use of LCP in implantable neural interfaces and are worthy of following in other biomedical application.

Seoul, Korea (Republic of)   Prof. Sung June Kim
February 2018

Abstract

Biocompatible polymers have gained widespread interest for implantable biomedical applications owing to their flexibility and compatibility with micro-fabrication processes. A liquid crystal polymer (LCP) is an inert, highly water-resistant, and thermoplastic polymer suitable for the encapsulation of electronic components and as a base material for fabricating neural interfaces. LCP-based implantable devices have salient benefits in terms of performance and reliability owing to their extremely low water absorption rate (<0.04%) and applicability for monolithic integration in neural interfaces and electronics packaging. In this dissertation, a new design for LCP-based neural interfaces especially for cochlear electrode arrays is proposed and evaluated in terms of its fabrication process, functionality, and reliability. The following issues of an LCP-based cochlear electrode arrays were studied: atraumatic deep insertion, tripolar stimulation, and long-term reliability.

Flexible LCP-based cochlear electrode arrays have been studied, but no electrode structure has been designed for atraumatic insertion. An atraumatic cochlear electrode array has become essential for high-performance cochlear implants, such as electric acoustic stimulation (EAS), where the preservation of residual hearing is important. A new design for an LCP-based cochlear electrode array for atraumatic implantation that uses precise batch fabrication and a thermal lamination process is proposed, which is unlike conventional wire-based cochlear electrode arrays. Multilayered structure with variable layers of LCP films depending on the parts of the array was designed to achieve a sufficient degree of basal rigidity while maintaining a flexible tip, and a peripheral blind via was used to reduce the width of the array. The resultant electrode array was 0.3 and 0.75 mm in diameter at the tip and base, respectively. In vitro force measurements in a customized experimental setup revealed that the insertion force (with a displacement of 8 mm from a round window) and the maximum extraction force are 2.4 and 34.0 mN, respectively. Five human temporal bone insertion trials showed that the electrode arrays can be inserted from 360 to 630° without trauma at the basal turn. Electrically evoked auditory brainstem responses were successfully recorded in a guinea pig model, which confirms the efficacy of the array. Hearing preservation and tissue reaction were investigated during implantation of the LCP electrode array and 4 weeks afterward.

Channel interaction is an important consideration for high-density, pitch-recognizing cochlear implants. There have been efforts to increase distinct stimulation channels using advanced focused current stimulation methods, including tripolar stimulation. In this dissertation, structural considerations on electrode sites are discussed for locally focused stimulation. A three-dimensional (3D) arrangement of electrode site in multilayered structure can be fabricated because differently patterned LCP layers can be merged into one substrate using thermal compression bonding. The 3D electrode site structures for locally tripolar stimulation are simulated about an electrical field distribution using the finite element method. An LCP electrode array of center stimulation channels with sidewall auxiliary channels for tripolar stimulation is fabricated based on the result of simulation. Compared to conventional monopolar and tripolar stimulation, locally tripolar stimulation on the proposed electrode site structure is more focused through in vitro measurements, which show the spreading of electrical stimulation in electrolytes.

Device reliability is one of the most significant issues in polymer-based neural prostheses. Two technical strategies are suggested in this dissertation. One strategy adopts a mechanical interlocking structure at the metal–polymer interface, which was started by J. H. Kim. This study extends his work and analyzes the impact of the strategy in terms of device reliability. The polymer–metal interface is vulnerable to water penetration that causes device failure. The goal is to suggest a feasible fabrication method using mechanical interlocking to improve polymer–metal adhesion in polymer-based neural electrodes and evaluate its impact on device reliability quantitatively through in vitro measurements. After the metal patterns with undercut profile cross sections are fabricated using a dual photolithography process and electroplating, the LCP interlocks with the metal during the lamination process. In a 180° peel test, the average maximum adhesion force of the samples with and without mechanical interlocking was 19.24 N and 14.27 N, respectively. In vitro accelerated soak tests that consist of interdigitated electrode patterns and a customized system for measuring the leakage current show that samples with and without interlocking fail to function after 224 days and 185 days, respectively, in a 75 °C saline solution. Scanning electron microscopy revealed that the interlocked LCP–metal interfaces remained intact after water leakage.

The other strategy is to use dielectric materials in LCP-based neural implants. Dielectric materials, such as silicon dioxide and silicon nitride, have been used in neural implants to prevent water and ion penetration. In addition to these features, dielectric materials can maintain metal patterning during the lamination bonding process, which causes migration of metal patterning on the LCP substrate. Preliminary tests—including a peel test to compare LCP–dielectric interface adhesion strength to that of an LCP–LCP interface, and thermo-compression bonding of LCP and dielectric materials with metal patterning to observe metal migration—were performed with consideration to the role of dielectric materials in the LCP-based device and their effects on device reliability. The LCP–dielectric interface is more adhesive than the weakly bonded LCP–LCP interface, and there is no metal migration after the lamination process (295 °C, 1 MPa). The results confirm the feasibility of the strategy.

Finally, a review of the long-term reliability of LCP-based neural prosthetic devices, including recently developed enabling technologies, demonstrated prototype devices, their performance capabilities, and theoretical fundamentals, is presented. Verification shows the possibility of the development of cochlear electrode arrays for atraumatic deep insertion, advanced stimulation, and long-term clinical implants.

Keywords: Polymer-based neural prosthesis · Cochlear electrode array · Liquid crystal polymer · Atraumatic insertion · Focused stimulation · Long-term reliability

Student Number: 2012-20740

Parts of this thesis have been published in the following articles:

Some parts of this dissertation are extracted and adapted from the journal publications which were published during the course of this study:

- Tae Mok Gwon, Kyou Sik Min, Jin Ho Kim, Seung Ha Oh, Ho Sun Lee, Min-Hyun Park, and Sung June Kim, "Fabrication and evaluation of an improved polymer-based cochlear electrode array for atraumatic insertion," *Biomedical Microdevices*, 17(2), 2015.
- Tae Mok Gwon, Jin Ho Kim, Gwang Jin Choi, and Sung June Kim, "Mechanical interlocking to improve metal-polymer adhesion in polymer-based neural electrodes and its impact on device reliability," *Journal of Materials Science*, 51(14), pp. 6897–6912, 2016.
- Tae Mok Gwon, Chaebin Kim, Soowon Shin, Jeong Hoan park, Jin Ho Kim, and Sung June Kim, "Liquid crystal polymer (LCP)-based neural prosthetic devices," *Biomedical Engineering Letters*, 6(3), 2016.

Acknowledgements

I would like to thank my supervisor, Prof. Sung June Kim, for his guidance, mentorship, and advice. He has encouraged and inspired me with the intriguing research subjects. He has taught researcher's mental attitude to me since I first entered laboratory. I have learned that sincerity, modesty, and truthfulness are the most important virtues which researchers always have to consider. I will have kept the values in my mind by the end of my careers as a researcher.

I would like to express my gratitude to my family: father, mother, sister, and grandmother. They always cheer me up. Their support helps me to live a stable life, which makes me concentrate my attention on the researches for Ph.D. degree. I am very lucky and happy guy to be with them as a family. Also, I am appreciative of the cheers sent to me by my wife's family.

I am grateful to my co-workers and other members in our laboratory, NanoBio electronics and Systems laboratory (NBS) at Seoul National University. There was a plenty of discussion about the researches during group meetings and presentations. Their advices and helps have been significant impact on my thesis. Special thanks go to Dr. Kyou Sik Min and Dr. Jin Ho Kim for initiating and guiding the opportunities of the interesting research projects. Also, I would like to thank Inter-university Semiconductor Research Center (ISRC) where I spent many hours to realize my research projects.

Foremost, I would like to thank my co-worker, senior laboratory member, and wife, Soowon Shin for her kind help about research and comfort throughout my degree term. Without her support, encouragement, and love, I would have had more troubles during the experiments and not overcome difficult situations successfully. Ph.D. degree, good research project, and bright future career are valuable because she is with me. She makes me want to be a better man.

Contents

1 Introduction 1
  1.1 Overview of Neural Prostheses and Cochlear Implants 1
  1.2 Review of Cochlear Electrode Arrays 3
    1.2.1 Conventional Cochlear Electrode Arrays 3
    1.2.2 Polymer-Based Cochlear Electrode Arrays 4
  1.3 Proposed Polymer Cochlear Electrode Array 5
    1.3.1 Electrode Arrays for Atraumatic Deep Insertion 5
    1.3.2 Electrode Arrays for Tripolar Stimulation 5
  1.4 Long-Term Reliability of Polymer-Based Neural Prostheses 6
  1.5 Objectives of the Dissertation 7
  References 8

2 Materials and Methods 13
  2.1 Liquid Crystal Polymer (LCP) 13
    2.1.1 Material Properties and Types of LCP 15
    2.1.2 MEMS Technologies Compatible with LCP 16
  2.2 Cochlear Electrode Array for Atraumatic Deep Insertion 17
    2.2.1 Electrode Design 17
    2.2.2 Fabrication Process 20
    2.2.3 Experimental Setup and Protocol of in Vitro and
        in Vivo Evaluation Tests 21
  2.3 Polymer Electrode Array for Tripolar Stimulation 24
    2.3.1 Modeling and Simulation of Polymer-Based Cochlear
        Electrode Array for Tripolar Stimulation 24
    2.3.2 Fabrication Process 28
    2.3.3 In Vitro Measurements 29
  2.4 Long-Term Reliability Analysis of LCP-Based Neural
    Implants 30
    2.4.1 Overview of the Long-Term Reliability 31

            2.4.2   Technical Strategies to Improve Reliability of
                    LCP-Based Implantable Device   33
      References   40

3   **Results**   43
    3.1 LCP-Based Cochlear Electrode Array for Atraumatic Deep
         Insertion.   43
         3.1.1   Fabricated Electrode Array   43
         3.1.2   Insertion and Extraction Force Measurements   43
         3.1.3   Insertion Trauma in Human Temporal Bone Insertion
                 Study   44
         3.1.4   Electrically Evoked Auditory Brainstem Response
                 Recording   45
         3.1.5   Histological Change and Hearing Preservation.   45
    3.2 Polymer Electrode Array for Tripolar Stimulation.   47
         3.2.1   Simulation Results According to Electrode Site Design . . .   47
         3.2.2   Fabricated Electrode Array   48
         3.2.3   In Vitro Measurements.   49
    3.3 Long-Term Device Reliability   51
         3.3.1   LCP-Based Neural Electrode Array Using Mechanical
                 Interlocking at Metal-LCP Interface   51
         3.3.2   Fabrication Method Using LCP and Dielectric
                 Materials.   58

4   **Discussion**   65
    4.1 LCP-Based Cochlear Electrode Arrays for Atraumatic Deep
         Insertion.   65
         4.1.1   Comparison of the Current Proposed Electrode
                 Array to the Previous Electrode Array   65
         4.1.2   Improving Electrode Design Related to Insertion
                 Depth and Trauma.   67
         4.1.3   Aspects to Improve in the Fabrication Process.   69
         4.1.4   In Vivo Implantation   71
    4.2 Power Consumption and Stimulation Threshold of Tripolar
         Stimulation.   73
    4.3 Technical Strategies to Improve Device Reliability   74
         4.3.1   Mechanical Interlocking at the Metal-Polymer Interface . . .   74
         4.3.2   Hybrid Device Based on Polymer and Dielectric
                 Materials.   77
    4.4 Review of Long-Term Reliability of LCP-Based Devices   80
      References   84

5   **Conclusion**   87

Abbreviations

ABR   Auditory Brainstem Response
ASIC   Application Specific Integrated Circuit
CSCc   Cathodic Charge Storage Capacity
CVD   Chemical Vapor Deposition
DAC   Digital to Analog Converter
DBS   Deep Brain Stimulation
EABR   Electrically evoked Auditory Brainstem Response
EAS   Electric Acoustic Stimulation
EIS   Electrochemical Impedance Spectroscopy
FEM   Finite Element Method
FIB   Focused Ion Beam
FPC   Flexible Printed Circuit
ICP   Inductively Coupled Plasma
IDE   Interdigitated Electrode
ISO   International Organization for Standardization
LCP   Liquid Crystal Polymer
MEMS   Microelectromechanical System
MRI   Magnetic Resonance Imaging
MTTF   Mean Time To Failure
PBS   Phosphate Buffered Saline
PCB   Printed Circuit Board
PFOA   Perfluorooctanoic Acid
PTFE   Polytetrafluoroethylene
PWM   Pulse Width Modulation
RF   Radio Frequency
SEM   Scanning Electron Microscopy
SPL   Sound Pressure Level
TEM   Transmission Electron Microscopy

List of Figures

Fig. 2.1   Conceptual view of LCP-based implantable neural
            prosthesis 14
Fig. 2.2   **a** LCP film (Kuraray, Tokyo, Japan). **b** Optical image of LCP
            surface 16
Fig. 2.3   **a** Basic fabrication process of LCP-based neural implants
            using MEMS technologies. **b** Results of photolithography
            (i), electroplating and seed layer etching (ii). 17
Fig. 2.4   Cross-section view of the design of tapered LCP-based
            cochlear electrode array for atraumatic deep insertion. 19
Fig. 2.5   Image of proposed cochlear electrode array using blind via 19
Fig. 2.6   Fabrication process flow of tapered LCP-based cochlear
            electrode array using MEMS technologies, thin-film process,
            and thermal lamination process. **a** Seed layer (Titanium and
            gold) deposition. **b** Photolithography using AZ 4620
            photoresist. **c** Electroplating for thick metal lines.
            d Developing photoresist and wet etching seed layer using
            nitrohydrochloric acid and hydrofluoric acid. **e** Laser cutting of
            align holes and aligning LCP layers with pre-cut cover layer.
            f Thermal lamination process for monolithic encapsulation 21
Fig. 2.7   3D slicer version 4 23
Fig. 2.8   Acute implantation of the LCP-based cochlear electrode array
            into guinea pig. **a** ABR recording in the soundproof room.
            b Implantation surgery 23
Fig. 2.9   Dummy LCP electrode array implantation surgery for hearing
            preservation and histological evaluation 25
Fig. 2.10   Explanation of current focusing stimulation method 26
Fig. 2.11   Modeling of human cochlea and implantation of an electrode
            array 27
Fig. 2.12   Designs of electrode sites in the simulation 28
Fig. 2.13   Outline of the test board used in vitro measurements of current
            focusing 29

Fig. 2.14   **a** Block diagram and **b** Image of the experimental setup for
            in vitro current focusing measurements.   30
Fig. 2.15   Water penetration in **a** Metal-based neural prosthesis and
            b LCP-based neural prosthesis   31
Fig. 2.16   Three pathways through which water ingress occurs   32
Fig. 2.17   Conceptual view of mechanical interlocking at metal-polymer
            interface in the polymer-based neural electrode array   34
Fig. 2.18   Fabrication processes of a LCP-based neural electrode with
            mechanical interlocking using MEMS technologies, including
            oxygen plasma, e-gun evaporation, dual steps of
            photolithography and electroplating, and wet etching   35
Fig. 2.19   Design of test samples categorized according to the formation
            of the mechanical interlocking structures   36
Fig. 2.20   Experimental setup for in vitro accelerated soak test. **a** Test
            samples are immersed in PBS at 75 °C to imitate a body
            environment. The test setup consists of a multichannel current
            stimulator to apply continuous biphasic pulses, a picoammeter
            for leakage current measurements, an interfacing circuit to
            switch between the stimulation mode and the leakage current
            measurement mode, and a deionized water tank to maintain the
            ion concentration of the PBS. **b** The interfacing circuit board,
            with the switch for mode conversion, and the test electrode
            sample are connected with an extension cable and a board   37
Fig. 2.21   Migration of metal pattern after lamination process   39
Fig. 2.22   Experimental setup of customized peel test.   40
Fig. 3.1   LCP-based cochlear electrode array with tapered structure   44
Fig. 3.2   Insertion force measured in the plastic scala tympani model. . . .   44
Fig. 3.3   Extraction force measured in the plastic scala tympani
            model   45
Fig. 3.4   Human temporal bone insertion study of electrode 1 (Round
            window approach). **a** Micro-CT image for insertion depth
            (450°). **b** Reconstructed 3-D image. **c** Cross-sectional views of
            cochlea where LCP electrode array is inserted (Trauma 0)   46
Fig. 3.5   Human temporal bone insertion study of electrode 2 (Round
            window approach). **a** Micro-CT image for insertion depth
            (450°). **b** Reconstructed 3-D image. **c** Cross-sectional views
            (Trauma 0)   47
Fig. 3.6   Human temporal bone insertion study of electrode 3 (Round
            window approach). **a** Micro-CT image for insertion depth
            (360°). **b** Reconstructed 3-D image. **c** Cross-sectional views
            (Trauma 0)   48
Fig. 3.7   Human temporal bone insertion study of electrode 4 (Round
            window approach). **a** Micro-CT image for insertion depth
            (630°). **b** Cross-sectional views (Trauma 3)   49

Fig. 3.8   Human temporal bone insertion study of electrode 5
             (Cochleostomy approach). **a** Micro-CT image for insertion
             depth (495°). **b** Cross-sectional views (Trauma 0)   50
Fig. 3.9   ABR recording which indicates the threshold level
             at 35 dB SPL   50
Fig. 3.10   Recorded EABR with Stimulation **a** On and **b** Off.   51
Fig. 3.11   ABR threshold shift after **a** 4 days, **b** 1 week, **c** 2 weeks,
             d 3 weeks, and **e** 4 weeks implantation   52
Fig. 3.12   **a** Enucleated cochlea after 4 weeks implantation.
             b Cross-section of the cochlea with LCP electrode
             implantation   53
Fig. 3.13   Cross-section view of stained cochlea in **a** Middle
             and **b** Basal turn   53
Fig. 3.14   Simulation results of tripolar stimulation according to
             electrode designs. **a** Design 1-1, **b** Design 1-2, **c** Design 2,
             d Design 3-1, **e** Design 3-2, **f** Design 3-3 [(i) electrode
             structure, (ii) absolute value of e-field at the plane of
             analyzing, (iii) picture of vector distribution of e-field]   54
Fig. 3.15   Fabrication process of LCP cochlear electrode array for locally
             tripolar stimulation using MEMS technologies, laser
             micromachining, and thin-film processes   58
Fig. 3.16   Results of fabrication process. **a** Metal patterning after wet
             etching of seed layer. **b** Via opening using laser ablation.
             c Laser micromachining of site opening. **d** Opened electrode
             site showing center and side electrodes.   59
Fig. 3.17   PWM data and biphasic pulses generated by ASIC chip and
             test board   60
Fig. 3.18   Optical images of Electrode site and IDE pattern in fabricated
             electrode array with and without mechanical interlocking
             pattern.   60
Fig. 3.19   SEM images of fabricated electrode array with mechanical
             interlocking. **a** Electrode site opening. **b** Seamless lamination
             result of upper and substrate layer of LCP. **c** Mechanically
             interlocked LCP film with undercut metal pattern.   61
Fig. 3.20   Impedance measurements of fabricated test samples with and
             without mechanical interlocking used in accelerated
             soak test   62
Fig. 3.21   Cyclic voltammogram to calculate charge storage capacitance
             (cathodic) of group 1 and group 2   62
Fig. 3.22   Leakage current measurements during the in vitro 75 °C
             accelerated soak tests   63
Fig. 3.23   Metal patterning **a** Before and **b** After lamination process
             using dielectric material and LCP.   63

Fig. 4.1   Comparison of LCP-based cochlear electrode arrays between
            the previous and the current designs. **a** The thickness of the
            previous LCP structure is 50 μm along its entire length. Its
            diameters at the tip and at the base are 0.5 and 0.8 mm,
            respectively, and its length was 28 mm. **b** The thickness of the
            current LCP structure in the cochlear electrode array varies
            from 25 μm (tip) to 75 μm (base). **c** Previously reported array
            employed a design scheme which arranges stimulation sites at
            the center with the lead wires around the stimulation sites. **d** In
            this design, a peripheral via is employed so as to protect the
            lead wires at the center from the cutting laser beam   66
Fig. 4.2   **a** Optical images of fabricated LCP-based cochlear electrode
            arrays with and without tapered structure. **b** Side and
            cross-sectional views of the arrays with and without peripheral
            blind via   68
Fig. 4.3   Insertion force measured in a customized insertion setup
            (25 and 50 μm tip mean the current design and the previous
            design, respectively)   69
Fig. 4.4   Extraction force measurements   70
Fig. 4.5   Insertion depth and trauma in human temporal bone insertion
            studies of LCP-based cochlear electrode arrays   71
Fig. 4.6   Fabrication processes and their results   72
Fig. 4.7   Deformation of metal line after lamination process.   72
Fig. 4.8   Comparison of fibrosis between conventional and LCP-based
            cochlear electrode array   73
Fig. 4.9   Scheme of electrical pulse trains of sequential and paired
            stimulation   74
Fig. 4.10   Optical microscope images of test samples. **a** Electrode site of
            test sample without mechanical interlocking in the course of
            accelerated soaking. **b** Electrode site of test sample with
            mechanical interlocking after water penetration. **c, d** Backlight
            microscope Images of test samples without and with
            mechanical interlocking after device failure. **e** Side view of
            delamination of LCP film. **f** Breakage of metal lines after water
            penetration   76
Fig. 4.11   SEM images of cross-section views of test samples after
            accelerated soak tests. **a** Without mechanical interlocking.
            b With mechanical interlocking   77
Fig. 4.12   Fabrication process of mechanically interlocked metal
            deposition onto LCP film (1st sacrificial layer and SiO_2
            deposition on Si or Glass wafer → Interlocking patterning
            using photolithography and SiO_2 etching → 2nd Sacrificial
            layer and noble metal deposition → Lamination metal

	patterning onto LCP film → SiO_2 and 2nd sacrificial layer removal)	78
Fig. 4.13	Mechanical interlocked metal-LCP in the substrate layer	78
Fig. 4.14	**a** TEM and **b** FIB-SEM images of micro-crack in electrode site.	79
Fig. 4.15	Images of LCP-LCP interface. **a** Using high-temperature, high-pressure lamination process and **b** Low-temperature, low-pressure lamination process after peeling off	79
Fig. 4.16	Fabrication flow feasible to make LCP-dielectric material hybrid implantable device.	80
Fig. 4.17	Electrical modeling of crosstalk analyzation in stimulation electrode when water penetration occurs.	82

List of Tables

Table 2.1 Properties of silicon, silica, and polymeric biocompatible materials. 15

Table 2.2 Conditions of current stimulation in the simulation 26

Table 2.3 Features of electrode site designs for locally tripolar stimulation 28

Table 3.1 Summarized results of absolute value of e-field distribution. 57

Table 3.2 Summarized results of relative ratio of e-field distribution 57

Table 3.3 Comparison of reduction ratio of voltage amplitude at 1 mm distant from the center electrode 59

Table 3.4 Results of peel tests in test samples with and without mechanical interlocking 61

Table 3.5 Results of customized peel tests to compare bonding strength of LCP-LCP with LCP-dielectric materials. 61

Table 4.1 Summarized results of reliability tests of LCP-based devices and the conditions of each specimen 83

Chapter 1
Introduction

1.1 Overview of Neural Prostheses and Cochlear Implants

Neural prostheses are electronic devices that can restore or substitute the partially damaged or profoundly impaired nervous system from neural diseases. Significant developments, from basic scientific discoveries to practical engineering technologies, have been achieved in many applications [1–4]. These findings and research combined with medical approaches have led to success in the commercialization of neural prostheses, such as cochlear implants, deep brain stimulation (DBS), and artificial retinas. Cochlear implantations have been performed in over 300,000 cases worldwide for people with impaired auditory function caused by hair cell damage [5–8]. DBS is a clinically qualified neuromodulation method for the effective alleviation of Parkinson's disease symptoms, essential tremors, and dystonia [9–11]. Visual prostheses, including artificial retinas (which replace the function of the retina) for people who have retinitis pigmentosa or age-related macular degeneration, have been manufactured and used as implantable devices in recent years [12, 13].

Neural implants consist of microelectrode arrays, microelectronics, and implantable microsystem packages. A neural microelectrode array is an interface that delivers electrical pulses and signals to neural tissue. Various materials have been used to fabricate microelectrode arrays, such as metal wires, silicon, polymers, and carbon fibers [1, 14–17]. Metal wire-based electrodes are one of the most powerful and reliable tools to both stimulate and record the nervous system [1]. Silicon-based neural probes are also widely used owing to their elaborate structure and precise size control (owing to the semiconductor fabrication process) [18, 19]. Carbon fibers and polymers have consistently attracted attention for their uses in implantable biomedical devices because of their flexibility, durability, and compatibility with the microfabrication process [20–22]. There are many types of neural electrode arrays because the electrode size, design, and stiffness depend on the target neural tissue and the biological anatomy where the electrode array is to be implanted. Therefore, the fundamental requirements for the neural electrode array are different in each application.

© Springer Nature Singapore Pte Ltd. 2018
T. M. Gwon, *A Polymer Cochlear Electrode Array: Atraumatic Deep Insertion, Tripolar Stimulation, and Long-Term Reliability*, Springer Theses,
https://doi.org/10.1007/978-981-13-0472-9_1

For example, neural depth probes, which are used to record neural signals in brain, must be stiff enough to insert into brain tissue, but non-breakable so they can be maintained in the tissue. Cochlear electrode arrays must be both insertable and flexible because the cochlea is a spiral structure. Cuff-type electrode arrays have been used in peripheral neural tissues to cover axil nerves without insertion or damage to the nerve tissues. On the other hand, microelectronics for stimulation, recording, and power-data transmission have been developed to precisely control neural responses and derive diverse effects in local spaces. Application-specific integrated circuit (ASIC) chips for neural stimulation, wireless transmission technology (including inductive links), Bluetooth and Zigbee devices, data transmission algorithm strategies, and signal processing strategies, have contributed to the success of high-performance neural implants. Implantable microsystem packages constitute the technology that enables electronics to be implanted in the human body. Without hermetic packages, neural implants could only be used for animal testing. Conventional implantable devices have been fabricated using metal- and ceramic-encased packaging technologies [23–26]. Recently, polymer-based packages have been actively studied owing to their compatibility with the microfabrication process and miniaturization, lightness, transparency to radio frequency (RF) waves, and cost effectiveness [27–33].

Cochlear implants are a class of neural prostheses that can perform the role of hair cells that convert mechanical energy to electrical signals in the cochlea. Movements of the stereocilia of the hair cells are detected and converted to electrical signals that pass through the upper auditory pathway. Cochlea implants mimic hair cells generating electrical pulses according to external sound signals and deliver the pulses directly to the next section of the auditory path past the hair cells, such as the spiral ganglion cells. Cochlear implants are one of the most successful commercial neural implantable devices. The internal device of a cochlear implant is composed of an intracochlear electrode array and an electronics package that includes a coil for data-power transmission, a current pulse generator, and a data receiver. The external device of a cochlear implant consists of a microphone for receiving sound signals, an analog to digital converter, a signal processor for data conversion, and a transmission coil.

Conventional cochlear implants have been manufactured by three major global companies: Cochlear Ltd., MED-EL, and Advanced Bionics. Most implantable devices implanted in human cochleae are based on metal-encased electronics packages and wire-assembled, platinum-iridium electrode arrays. The fabrication of the conventional devices is a laborious manual process, which leads to low throughput. Feedthrough for the connection between the electrode array and electronics package requires heterogeneous hermetic sealing, which is a complicated, expensive technology [34, 35]. Additionally, metal-encased packages block electromagnetic wave penetration, which prevents coil integration into the electronics package. Size reduction of the implantable unit is difficult because of the external coil. The number of stimulation channels of an intracochlear electrode array, which determine the maximum data and information quantity, is limited to the capabilities of the metal-wire fabrication process. There have been efforts to solve the problems conventional cochlear implants face by using new polymer materials and microfabrication processes compatible with mass-production [34, 36].

1.2 Review of Cochlear Electrode Arrays

Cochlear electrode arrays play an important role in delivering electrical pulses to neural tissues in the auditory pathway. Because electrode arrays contact neural tissue directly and are exposed to electrical stimuli and electrolytes, materials for implantable electrode arrays must be selected with consideration to biocompatibility, durability, charge storage capacity, encapsulation ability, and reliability. Moreover, electrode design must fit into the cochlea. Unlike other neural implant applications, the spiral structure of the cochlea is highly complicated. Human cochleae occupy three spaces: the scala tympani, scala vestibule, and scala media. Cochlear electrode arrays are inserted into the scala tympani, which meets the scala vestibule at the basilar membrane. Electrode arrays must be in the scala tympani during insertion to protect cochlear tissue. Insertion depth and the number of stimulation channels affect the perception ability of the recipients. Basal and apex turns of the cochlea respond to high and low frequency, respectively. Atraumatic and deep insertion can be advantageous for pitch perception and the noise recognition [37–39]. On the other hand, localized current stimulation enables high-performance cochlear implants. There have been various stimulation techniques designed to increase the number of distinguishable channels and reduce channel interaction, which supposedly cause difficulties in pitch perception and noise recognition [40–43].

1.2.1 Conventional Cochlear Electrode Arrays

Conventional cochlear electrode arrays are based on Teflon-insulated Pt–Ir wire and silicone elastomers. Manual processes of aligning wire electrodes and making ball- or band-type electrode sites are laborious, low-throughput, and expensive [44]. The number of stimulation channels ranges is 12–26 [45]. The length of the electrode arrays is 19–31.5 mm. Short electrode arrays are used for electrical acoustic stimulation systems that stimulates the low-frequency region via sound signal and the high-frequency region via electrical pulses. The diameter of the basal end is 0.8–1.3 mm. Flexible array tips are essential to minimize cochlear electrode array insertion trauma.

Electrode shape has been studied and developed for high performance and minimal insertion trauma [46, 47]. Pre-curved (perimodiolar) electrode arrays with insertion tools have been developed to be close to the target neuron cells, such as spiral ganglion cells. This modiolus-hugging structural electrode array showed a lower threshold than the straight electrode array owing to the distance between the stimulation electrode site and the target tissue [48, 49].

1.2.2   Polymer-Based Cochlear Electrode Arrays

Biocompatible polymers, such as polyimide, parylene-C, silicone elastomer, and
SU-8, have attracted attention as microelectrode array substrate materials owing to
their flexibility and feasibility for microfabrication (thin-film fabrication). Among
these biocompatible polymers, liquid crystal polymers (LCPs) popular biocompatible
polymers for biomedical implantable devices. They are one of the most reliable
polymers in water because their water absorption rate is less than 0.04%. LCP-
based devices allow for the monolithic integration of electrode arrays and electronic
packages for enhanced reliability. LCP-based electrode arrays are non-breakable
because these arrays simply consist of flexible polymers and metal patterns. Cochlear
electrode arrays using LCPs have been studied by several groups. The first LCP-based
cochlear electrode array was fabricated in 2006 using 50-μm-thick LCP films with
a high melting temperature, on which metal is deposited [50]. Films with a lower
melting temperature were used as a bonding layer during the thermal lamination
process. The length of the electrode array was less than 30 mm, and there were 12
contact points. The apical dimensions of the array were about 250 μm × 250 μm
(width × height). Medical-grade silicone was adopted as an overmold material for
better positioning of the electrode array near the cochlea. It was found in an acute
animal insertion trial that it was too stiff for clinical use with humans. Next, improved
highly-flexible, LCP-based cochlear electrode arrays were fabricated and evaluated
[51]. Only two 25-μm-thick LCP layers with a low melting temperature for metal
patterning and bonding were used, resulting in a 50-μm-thick electrode. Medical-
grade silicone elastomer (50-μm-thick) was used to encapsulate its entire length using
a novel self-aligned molding method. Its diameter at the tip and base was 0.5 and
0.8 mm, respectively, and its length was 28 mm. The electrochemical properties
showed that it could be used as a stimulation electrode, and the insertion force was
half that of a comparative wire-based electrode array. The results of human temporal
bone insertion studies revealed insertion trauma and feasibility of its clinical use.

There have been batch-fabricated cochlear electrode arrays using bulk micromachining and lithographically-defined thin-film technology [52–54]. Silicon-based,
hybrid, thin-film cochlear electrode arrays were encapsulated using Parylene-C and
integrated with position sensors. Silicon-based and LCP-based cochlear electrode
arrays offer a higher density of electrode sites than conventional cochlear electrode
arrays owing to the use of the precise thin-film and less manual processes. Position
sensors were adopted to minimize insertion damage.

1.3   Proposed Polymer Cochlear Electrode Array

The issues of the cochlear electrode array under consideration in this dissertation are as follows:

(1) Compatibility with micro-fabrication processes for mass production
(2) Atraumatic insertion for hearing preservation
(3) Low-frequency region stimulation
(4) High-density electrode arrays and reduction of channel interaction between electrode sites for locally distinct stimulation

Polymer cochlear electrode arrays have been studied as a possible alternative to conventional cochlear electrode arrays because of these issues. Cochlear electrode arrays based on thin-film polymers can be compatible with microfabrication processes, such as semiconductor fabrication processes, monolithic encapsulation, and microelectrodemechanical systems (MEMS). A differentiated cochlear electrode array design is suggested, fabricated, and evaluated for atraumatic deep insertion. Additionally, a novel electrode structure, which is enabled in the multi-layered thin-film electrode array, is proposed and evaluated to verify the feasibility of localized current stimulation for reducing channel interaction.

1.3.1   Electrode Arrays for Atraumatic Deep Insertion

Atraumatic electrode insertion and a stable interface between the electrode array and neural tissues are essential for high performance cochlear implants because atraumatic electrode insertion is important for hearing preservation, especially for electric acoustic stimulation (EAS), which uses both high-frequency electrical stimulation and low-frequency sound stimulation. Deep insertion is advantageous for the perception of music and recognition of noise [38, 39].

In this study, an LCP-based cochlear electrode array for atraumatic deep insertion is proposed. An electrode array compatible with precise batch processing, thin-film processes, and a thermal lamination process, is proposed. The differentiated design of the proposed LCP-based cochlear electrode array adopts a multi-layered structure with layers of LCP films that achieve a sufficient degree of basal stiffness and tip flexibility. The design also uses a shadow masking method with a blind via to reduce the width of the electrode array, resulting in a finer electrode array than previous arrays.

1.3.2   Electrode Arrays for Tripolar Stimulation

Limitations on the number of stimulation channels in current cochlear implants come from not only the manual wire-based electrode array fabrication process, but also channel interaction, which leads to insignificant effects when there are more than eight stimulation channels [55]. Low spectral resolution could affect pitch perception and hearing in noisy environments. Current steering and focusing stimulation methods have been studied to increase the number of distinguishable perceptual channels, especially current-focusing, multi-polar current stimulation techniques to prevent the spread of electrical fields [40, 42, 43, 56]. Tripolar stimulation, which uses electrode channels adjacent to both sides of the stimulation channel as a reference for current pulses, has been shown to reduce current spread and increase spectral resolution [43]. Although there is controversy about the effect of tripolar stimulation in pitch perception, some reported spatial selectivity helps analyze the tuning position and improve cochlear implant performance [57, 58].

In this study, a novel electrode structure that can enhance the spatial resolution of tripolar stimulation is proposed. This structure is feasible for a multi-layered thin-film electrode array. Previously, tripolar stimulation was performed at three stimulation channels, but in the proposed strategy, local tripolar stimulation can be performed on one electrode channel using another layer of auxiliary electrode sites for tripolar stimulation. In other words, stimulation channels and auxiliary tripolar stimulation channels are separated in different LCP layers. A computer simulation is used to verify the possibility of the proposed electrode structure and the field distribution of the various designs. After choosing the best electrode design, a laminated LCP-based electrode array with laser ablation is fabricated.

1.4   Long-Term Reliability of Polymer-Based Neural Prostheses

Long-term reliability is one of the major issues in polymer-based neural prostheses. Polymer-based neural implants are more vulnerable to electrolyte ingress than conventional metal-encased and ceramic-based packages owing to their non-hermeticity and high permeability to water and gas. Device reliability in polymer-based neural implants is difficult to measure using conventional helium leak tests, which is an industry standard method to quantify the hermeticity of metal or ceramic cases, because polymer has a permeable surface where the helium is absorbed during bombing. The absorbed helium is gradually released, resulting in a misleading leak rate [59–62]. Because there is no practically accepted standard testing method to measure long-term reliability of polymer-based neural implants, LCP-based implantable devices have been evaluated using the helium fine leak test and accelerated soak test. The accelerated soak test is a testing method that measures device reliability in a wet environment that mimics the body fluidic conditions at an elevated temperature to estimate device lifetime in a shortened time. The estimation of device lifetime at body temperature is based on the Arrhenius equation using the calculated reaction rate and the mean time to failure at the accelerated temperature. Moreover, in vivo device stability experiments and in vitro reliability tests have also been conducted [63].

Delamination in polymer-based devices has a detrimental effect on device reliability. Most polymer-based devices suffer from delamination by water ingress into
the surface of the polymer and the weakly bonded interfaces. Therefore, interface
adhesion should be enhanced using mechanical or chemical modification. There have
been many studies on long-term reliability of implantable polymer-based devices and
methods to enhance device reliability. For polyimide, effective annealing steps to prevent delamination were developed [64, 65]. For parylene-C, chemical modification,
adhesion promoter, and structural approaches to increase adhesion reliability were
used at the parylene–metal interface [66, 67].

In this dissertation, the long-term reliability of LCP-based neural implants is analyzed and reviewed. Failure mechanisms of LCP-based thin-film neural implants
are discussed. Two technical strategies to improve the reliability of LCP-based
implantable devices are suggested and evaluated using in vitro measurements. Strategy 1 mentions mechanical interlocking to strengthen metal-LCP adhesion, which
is the most vulnerable part to water penetration, as stated by Kim [68]. This study
extends his work and analyzes the impact of the device reliability strategy. Metal patterning onto electrode sites using dual photolithography and electroplating interlocks
with the LCP-insulation layer during the thermal lamination process are discussed.
In vitro adhesion and accelerated soak tests are used to verify the impact of mechanical interlocking on device reliability. Strategy 2 deals with the fabrication method
for microimplantable device using LCP and dielectric materials. To enhance device
reliability, dielectric materials are introduced to prevent water penetration and metal
migration, and a hybrid structure is proposed. The feasibility of the proposed method
is investigated using preliminary tests, including a peel test and a high temperature
and pressure thermos-compression bonding process using a hybrid LCP-dielectric
material structure

1.5 Objectives of the Dissertation

In this study, a tapered, multi-layered, LCP-based cochlear electrode array for atraumatic deep insertion is suggested. Metal lines and electrode sites are patterned on a
flexible LCP film using MEMS technologies and a thin-film process. The electrode
array is designed with variable thermo-compression laminated layers and peripheral blind vias. Employing the proposed design scheme, the width of the electrode
array becomes smaller than that of previously reported LCP-based cochlear electrode
arrays. Consequently, the possibility of insertion trauma is lower. The measurements
to prove atraumatic insertion are presented. The insertion forces and extraction forces
in the scala tympani model are measured and compared to conventional cochlear electrode arrays. Insertion trauma and depth are investigated through human temporal
bone insertion studies.

A local tripolar stimulation method using a multi-layered electrode structure is
also proposed in this dissertation. Simulation results that enlarge the effect of current
focusing to reduce current spread are presented. A fabrication process for multi-

layered local tripolar stimulation electrode arrays is devised, and fabricated electrode arrays are evaluated through in vitro measurements using a test board based on ASIC chips for current pulse generation.

Long-term reliability of LCP-based neural implants is analyzed and discussed. The failure mechanism and suggestions to improve the long-term reliability of LCP-based implantable devices are presented. Adopting MEMS technologies, mechanical interlocking patterns on the electrode site for the enhancement of metal-LCP adhesion and fabrication methods using LCPs and dielectric materials are suggested and evaluated. Results of peel tests and in vitro accelerated soak tests show that the device reliability can be improved. Long-term reliability of LCP-based neural implantable devices is reviewed and the quest for desirable ways to enhance reliability is described.

The specific objectives of the dissertation are as follows:

(1) Fabrication and evaluation of an LCP-based cochlear electrode array for atraumatic deep insertion
(2) Suggestion and feasibility of an LCP-based cochlear electrode array for tripolar stimulation
(3) Suggestion and evaluation of technical strategies to enhance long-term reliability of LCP-based neural implants: mechanical interlocking to strengthen metal-LCP adhesion and LCP-dielectric material hybrid fabrication methods
(4) Review and analysis of the reliability of LCP-based implantable devices.

References

1. S.F. Cogan, Neural stimulation and recording electrodes. Annu. Rev. Biomed. Eng. **10**, 275–309 (2008)
2. V.K. Khanna, "Biomaterials for Implants," in *Implantable Medical Electronics: Prosthetics, Drug Delivery, and Health Monitoring* (Springer International Publishing, Cham, 2016) pp. 153–166
3. T. Stieglitz, M. Schuetter, K.P. Koch, Implantable biomedical microsystems for neural prostheses. IEEE Eng. Med. Biol. Mag. **24**, 58–65 (2005)
4. A. Prochazka, V.K. Mushahwar, D.B. McCreery, Neural prostheses. J. Physiol. **533**, 99–109 (2001)
5. B.J. Gantz, B.F. Mccabe, R.S. Tyler, Use of multichannel cochlear implants in obstructed and obliterated cochleas. Otolaryngol.-Head Neck Surg. **98**, 72–81 (1988)
6. K.D. Brown, T.J. Balkany, Benefits of bilateral cochlear implantation: a review. Curr. Opin. Otolaryngol. Head Neck Surg. **15**, 315–318 (2007)
7. B.S. Wilson, C.C. Finley, D.T. Lawson, R.D. Wolford, D.K. Eddington, W.M. Rabinowitz, Better speech recognition with cochlear implants. Nature **352**, 236–238 (1991)
8. F.-G. Zeng, S.J. Rebscher, Q.-J. Fu, H. Chen, X. Sun, L. Yin et al., Development and evaluation of the Nurotron 26-electrode cochlear implant system. Hear. Res. **322**, 188–199 (2015)
9. D.S. Kern, R. Kumar, Deep brain stimulation. The Neurologist **13**, 237–252 (2007)
10. H.S. Mayberg, A.M. Lozano, V. Voon, H.E. McNeely, D. Seminowicz, C. Hamani et al., Deep brain stimulation for treatment-resistant depression. Neuron **45**, 651–660 (2005)
11. J.S. Perlmutter, J.W. Mink, Deep brain stimulation. Annu. Rev. Neurosci. **29**, 229 (2006)

12. M.S. Humayun, J.D. Dorn, L. da Cruz, G. Dagnelie, J.A. Sahel, P.E. Stanga et al., Interim results from the international trial of second sight's visual prosthesis, Ophthalmology, **119** (2012)
13. R.K. Shepherd, M.N. Shivdasani, D.A.X. Nayagam, C.E. Williams, P.J. Blamey, Visual prostheses for the blind. Trends Biotechnol. **31**, 562–571 (2013)
14. M. HajjHassan, V. Chodavarapu, S. Musallam, NeuroMEMS: neural probe microtechnologies. Sensors **8**, 6704 (2008)
15. L. Grand, A. Pongrácz, É. Vázsonyi, G. Márton, D. Gubán, R. Fiáth et al., A novel multisite silicon probe for high quality laminar neural recordings. Sens. Actuators, A **166**, 14–21 (2011)
16. G. Grigori, E.M. Jeffrey, A.L. William, J.G. Timothy, A carbon-fiber electrode array for long-term neural recording. J. Neural Eng. **10**, 046016 (2013)
17. J.W. Thomas, C.W. Michael, L.M. Janice, L.P. Rachel, U.E. Jeremiah, Nano-biotechnology: carbon nanofibres as improved neural and orthopaedic implants. Nanotechnology **15**, 48 (2004)
18. K.C. Cheung, Implantable microscale neural interfaces. Biomed. Microdevice **9**, 923–938 (2007)
19. S. Herwik, S. Kisban, A.A.A. Aarts, K. Seidl, G. Girardeau, K. Benchenane et al., Fabrication technology for silicon-based microprobe arrays used in acute and sub-chronic neural recording. J. Micromech. Microeng. **19**, 074008 (2009)
20. D.-H. Kim, J.A. Wiler, D.J. Anderson, D.R. Kipke, D.C. Martin, Conducting polymers on hydrogel-coated neural electrode provide sensitive neural recordings in auditory cortex. Acta Biomater. **6**, 57–62 (2010)
21. V.S. Polikov, P.A. Tresco, W.M. Reichert, Response of brain tissue to chronically implanted neural electrodes. J. Neurosci. Methods **148**, 1–18 (2005)
22. M.D. Johnson, R.K. Franklin, M.D. Gibson, R.B. Brown, D.R. Kipke, Implantable microelectrode arrays for simultaneous electrophysiological and neurochemical recordings. J. Neurosci. Methods **174**, 62–70 (2008)
23. S.K. An, S.I. Park, S.B. Jun, C.J. Lee, K.M. Byun, J.H. Sung et al., Design for a simplified cochlear implant system. IEEE Trans. Biomed. Eng. **54**, 973–982 (2007)
24. P. Maló, M. de Araújo Nobre, J. Borges, R. Almeida, Retrievable metal ceramic implant-supported fixed prostheses with milled titanium frameworks and all-ceramic crowns: retrospective clinical study with up to 10 years of follow-up. J. Prosthodont. **21**, 256–264 (2012)
25. J.F. Patrick, P.A. Busby, P.J. Gibson, The development of the nucleus® freedom™ cochlear implant system. Trends in Amplification **10**, 175–200 (2006)
26. C.M. Zierhofer, I.J. Hochmair, E.S. Hochmair, The advanced Combi 40+ cochlear implant. Am. J. otol. **18**, S37–S38 (1997)
27. A.J.T. Teo, A. Mishra, I. Park, Y.-J. Kim, W.-T. Park, Y.-J. Yoon, Polymeric biomaterials for medical implants and devices. ACS Biomater. Sci. Engin. **2**, 454–472 (2016)
28. V. Castagnola, E. Descamps, A. Lecestre, L. Dahan, J. Remaud, L.G. Nowak et al., Parylene-based flexible neural probes with PEDOT coated surface for brain stimulation and recording. Biosens. Bioelectron. **67**, 450–457 (2015)
29. R.A. Normann, E.M. Maynard, P.J. Rousche, D.J. Warren, A neural interface for a cortical vision prosthesis. Vision. Res. **39**, 2577–2587 (1999)
30. F.J. Rodri, D. Ceballos, M. Schu, A. Valero, E. Valderrama, T. Stieglitz et al., Polyimide cuff electrodes for peripheral nerve stimulation. J. Neurosci. Methods **98**, 105–118 (2000)
31. L. Kee-Keun, H. Jiping, S. Amarjit, M. Stephen, E. Gholamreza, K. Bruce et al., Polyimide-based intracortical neural implant with improved structural stiffness. J. Micromech. Microeng. **14**, 32 (2004)
32. C. Hassler, T. Boretius, T. Stieglitz, Polymers for neural implants. J. Polym. Sci., Part B: Polym. Phys. **49**, 18–33 (2011)
33. B.J. Kim, E. Meng, Review of polymer MEMS micromachining. J. Micromech. Microeng. **26**, 013001 (2015)
34. J.H. Kim, K.S. Min, S.K. An, J.S. Jeong, S.B. Jun, M.H. Cho et al., Magnetic resonance imaging compatibility of the polymer-based cochlear implant. Clin. Exp. Otorhinolaryngol. **5**, S19–S23 (2012)

35. G. Jiang, D.D. Zhou, Technology advances and challenges in hermetic packaging for implantable medical devices, in *Implantable neural prostheses 2: techniques and engineering approaches*, ed. by D. Zhou, E. Greenbaum (Springer New York, New York, NY, 2010), pp. 27–61

36. T.M. Gwon, C. Kim, S. Shin, J.H. Park, J.H. Kim, S.J. Kim, Liquid crystal polymer (LCP)-based neural prosthetic devices. Biomed. Engin. Lett. **6**, 148–163 (2016)

37. S.J. Rebscher, A. Hetherington, B. Bonham, P. Wardrop, D. Whinney, P.A. Leake, Considerations for design of future cochlear implant electrode arrays: electrode array stiffness, size, and depth of insertion. J. Rehabil. Res. Dev. **45**, 731–747 (2008)

38. R. Shepherd, K. Verhoeven, J. Xu, F. Risi, J. Fallon, A. Wise, An improved cochlear implant electrode array for use in experimental studies. Hear. Res. **277**, 20–27 (2011)

39. P. Wardrop, D. Whinney, S.J. Rebscher, J.T. Roland Jr., W. Luxford, P.A. Leake, A temporal bone study of insertion trauma and intracochlear position of cochlear implant electrodes. I: comparison of nucleus banded and nucleus contour™ electrodes. Hear. Res. **203**, 54–67 (2005)

40. C.K. Berenstein, L.H.M. Mens, J.J.S. Mulder, F.J. Vanpoucke, Current steering and current focusing in cochlear implants: comparison of monopolar, tripolar, and virtual channel electrode configurations. Ear Hear. **29**, 250–260 (2008)

41. J.B. Firszt, D.B. Koch, M. Downing, L. Litvak, Current steering creates additional pitch percepts in adult cochlear implant recipients. Otol. Neurotology **28**, 629–636 (2007)

42. D.M. Landsberger, A.G. Srinivasan, Virtual channel discrimination is improved by current focusing in cochlear implant recipients. Hear. Res. **254**, 34–41 (2009)

43. B.H. Bonham, L.M. Litvak, Current focusing and steering: modeling, physiology, and psychophysics. Hear. Res. **242**, 141–153 (2008)

44. K.S. Min, S.B. Jun, Y.S. Lim, S.-I. Park, S.J. Kim, Modiolus-hugging intracochlear electrode array with shape memory alloy. Comput. Math. Methods Med. **2013**, 9 (2013)

45. Product brochure. *Electrode arrays: designed for atraumatic implantation providing superior hearing performance*. Available: http://s3.medel.com/pdf/21617.pdf

46. B.K. Chen, G.M. Clark, R. Jones, Evaluation of trajectories and contact pressures for the straight nucleus cochlear implant electrode array—a two-dimensional application of finite element analysis. Med. Eng. Phys. **25**, 141–147 (2003)

47. M. Tykocinski, L.T. Cohen, B.C. Pyman, T.J. Roland, C. Treaba, J. Palamara et al., Comparison of electrode position in the human cochlea using various perimodiolar electrode arrays. Otol. Neurotology **21**, 205–211 (2000)

48. E. Saunders, L. Cohen, A. Aschendorff, W. Shapiro, M. Knight, M. Stecker et al., Threshold, comfortable level and impedance changes as a function of electrode-modiolar distance. Ear Hear. **23**, 28S–40S (2002)

49. M.L. Hughes, P.J. Abbas, Electrophysiologic channel interaction, electrode pitch ranking, and behavioral threshold in straight versus perimodiolar cochlear implant electrode arrays. J. Acoust. Soc. Am. **119**, 1538–1547 (2006)

50. S. Corbett, J. Ketterl, T. Johnson, Polymer-based microelectrode arrays. MRS Online Proc. Libr. Arch. **926**, 0926 (2006). CC06-02 (6 pages)

51. K.S. Min, S.H. Oh, M.H. Park, J. Jeong, S.J. Kim, A polymer-based multichannel cochlear electrode array. Otol Neurotol **35**, 1179–1186 (2014)

52. P.T. Bhatti, A high-density thin-film electrode array for a cochlear prosthesis, Thesis University of Michigan, 2006

53. J. Wang, K.D. Wise, A hybrid electrode array with built-in position sensors for an implantable MEMS-Based cochlear prosthesis. J. Microelectromechan. Syst. **17**, 1187–1194 (2008)

54. J. Wang, K.D. Wise, A thin-film cochlear electrode array with integrated position sensing. J. Microelectromechan. Syst. **18**, 385–395 (2009)

55. L.M. Friesen, R.V. Shannon, D. Baskent, X. Wang, Speech recognition in noise as a function of the number of spectral channels: Comparison of acoustic hearing and cochlear implants. J. Acoust. Soc. Am. **110**, 1150–1163 (2001)

56. J.D. Falcone, P.T. Bhatti, Current steering and current focusing with a high-density intracochlear electrode array, in *Conference proceedings: Annual International Conference of the IEEE Engineering in Medicine and Biology Society. IEEE Engineering in Medicine and Biology Society* (2011), pp. 1049–52

57. C.A. Fielden, K. Kluk, C.M. McKay, Place specificity of monopolar and tripolar stimuli in cochlear implants: the influence of residual masking. J. Acoust. Soci. Am. **133**, 4109–4123 (2013)

58. Z. Zhu, Q. Tang, F.-G. Zeng, T. Guan, D. Ye, Cochlear-implant spatial selectivity with monopolar, bipolar and tripolar stimulation. Hear. Res. **283**, 45–58 (2012)

59. S. Costello, M.P.Y. Desmulliez, S. McCracken, Review of test methods used for the measurement of hermeticity in packages containing small cavities. IEEE Trans. Compon. Packag. Manuf. Technol. **2**, 430–438 (2012)

60. K. Aihara, M.J. Chen, C. Cheng, A.V.H. Pham, Reliability of liquid crystal polymer air cavity packaging. IEEE Trans. Compon. Packag. Manuf. Technol. **2**, 224–230 (2012)

61. A.-V. Pham, Packaging with liquid crystal polymer. IEEE Microwave Mag. **5**, 83–91 (2011)

62. B. Han, Measurements of true leak rates of MEMS packages. Sensors **12**, 3082–3104 (2012)

63. J. Jeong, S.H. Bae, J.-M. Seo, H. Chung, S.J. Kim, Long-term evaluation of a liquid crystal polymer (LCP)-based retinal prosthesis. J. Neural Eng. **13**, 025004 (2016)

64. J.S. Ordonez, C. Boehler, M. Schuettler, T. Stieglitz, Silicone rubber and thin-film polyimide for hybrid neural interfaces; A MEMS-based adhesion promotion technique, in *Neural Engineering (NER), 2013 6th International IEEE/EMBS Conference on* (2013), pp. 872–875

65. J.S. Ordonez, C. Boehler, M. Schuettler, and T. Stieglitz, Improved polyimide thin-film electrodes for neural implants, in *Engineering in Medicine and Biology Society (EMBC), 2012 Annual International Conference of the IEEE* (2012), pp. 5134–5137

66. J.H. C. Chang, L. Yang, K. Dongyang, T. Yu-Chong, Reliable packaging for parylene-based flexible retinal implant, in *2013 Transducers & Eurosensors XXVII: The 17th International Conference on Solid-State Sensors, Actuators and Microsystems (TRANSDUCERS & EUROSENSORS XXVII)* (2013), pp. 2612–2615

67. R. von Metzen, T. Stieglitz, The effects of annealing on mechanical, chemical, and physical properties and structural stability of Parylene C. Biomed. Microdevice **15**, 727–735 (2013)

68. J.H. Kim, A study on low-cost, effective, and reliable liquid crystal polymer-based cochlear implant system, Ph.D. Thesis, Department of electrical engineering and computer science, Seoul National University, 2015

Chapter 2
Materials and Methods

2.1 Liquid Crystal Polymer (LCP)

Biocompatible polymers such as polyimide, parylene-C, SU-8, and silicone elastomer have gained great interests in neural implants due to their flexibility and compatibility with micro-fabrication. In recent studies, polymer-based or hybrid with conventional neural implants have been developed and evaluated for clinical uses [1]. Polyimides have been widely used as insulation and substrate materials for medical devices. Polyimides can be used bulk film type or spun onto thin films as both photopatternable and non-photopatternable types. Polyimides are mainly served as substrates of flexible microelectrode arrays including cuff electrodes for peripheral nerve stimulation and micro-channel for nerve regeneration, and insulation materials for MEMS sensor. Moreover, high light transmittance property with wide range of wavelengths enable polyimide to be an optoelectronics material. Parylene-C has very low water absorption rate and chemical inertness. Chemical vapor deposition (CVD) process without any additives is needed to deposit parylene-C onto biomedical devices. It is mostly used as a coating material for implantable devices because a few micrometer of insulation layer is possible using parylene-C. SU-8 in MEMS have been famous due to its patternable property using photolithography and various types with usable thickness ranging from a few to hundreds of μm. Tunable electrical, mechanical, and optical properties make it more attractive biomaterials to be used in structural molds for soft lithography, optical waveguides, and neural probes.

Liquid crystal polymer (LCP) is another rising biocompatible material for biomedical devices. It has one of the lowest water absorption rate (<0.04%) and highest Young's Modulus among biocompatible polymers. Thermo-compression bonding and thermoforming technologies enable monolithic encapsulation of LCPs with variable designs, which makes it an attractive packaging material for implantable neural prosthetic devices [2]. MEMS technologies including metal deposition, photolithography, wet etching, dry (plasma) etching, and laser machining can be utilized on LCP film. As shown in Fig. 2.1, LCP is used not only for packaging

© Springer Nature Singapore Pte Ltd. 2018
T. M. Gwon, *A Polymer Cochlear Electrode Array: Atraumatic Deep Insertion, Tripolar Stimulation, and Long-Term Reliability*, Springer Theses,
https://doi.org/10.1007/978-981-13-0472-9_2

material, but also electrode substrate. Figure 2.1 shows conceptual view of LCP-based neural implants. LCP-based neural electrode array and LCP-based electronics packaging enable homogeneous adhesion at feedthroughs. Patterned metal lines and electrode sites on LCP substrate are insulated with another LCP layer and exposed by laser ablation [3]. LCP-based multichannel electrode arrays fabricated for stimulation or recording neuronal function can be monolithically encapsulated with the same LCP material for electronics packages [4]. This monolithic fabrication without feedthrough technologies enhances reliability of LCP-based neural implant. Thermal bonding and thermal forming process using mold with desired design allow tissue-conformable structure, which is a great advantage in relation to effectively functional implantation of devices. RF transparence of LCP and integrated coils onto LCP achieve miniaturized and wireless operated electronics packages [5]. In recent years, LCP-based microelectrode arrays for CI, artificial retina, DBS, electrocorticogram, and peripheral nerve stimulation have been developed and evaluated both in vitro and in vivo [4, 6–15]. Electrochemical and mechanical properties are enough to be used in clinical applications. Further, animal studies for efficacy of LCP-based microelectrode arrays and good magnetic resonance imaging (MRI) compatibilities and long-term evaluation tests show possibility of LCP-based implantable prostheses as substitutes of conventional neural implants [4, 16–19].

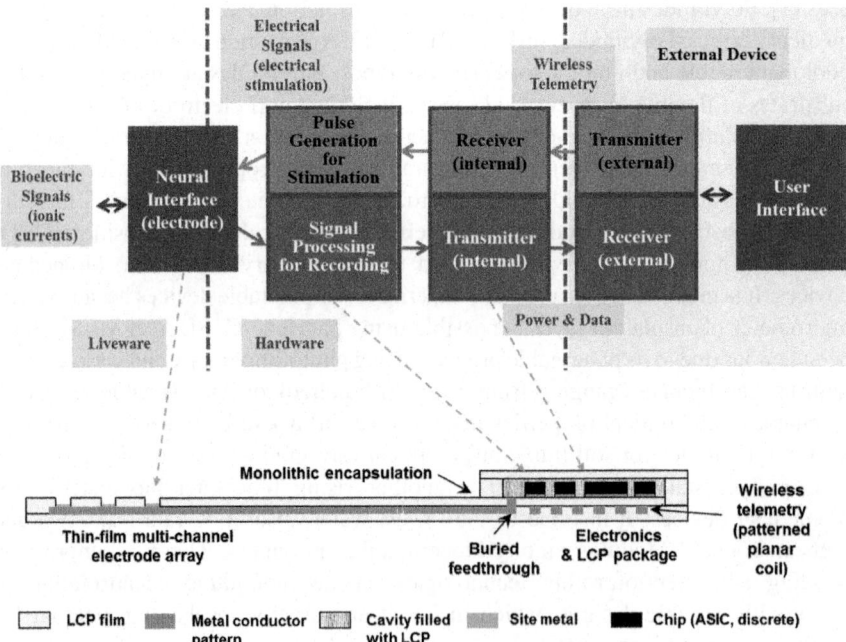

Fig. 2.1 Conceptual view of LCP-based implantable neural prosthesis

2.1.1 Material Properties and Types of LCP

LCP forms aligned molecule chains with a crystalline structure composed of rigid and flexible monomers. LCPs are distinguished by their mechanical, electrical, and optical properties because they are fabricated using chemically different base materials and processes for specific intended applications. The material properties of LCPs are summarized and compared with those of silicon, silica, and other biocompatible polymers in Table 2.1. Their Young's moduli range from 2 to 40 GPa. LCP is one of the most stable biocompatible polymers in water, making it a strong candidate substrate material for long-term implants. The melting points of LCPs range from 275 to 335 °C. It has an extremely low water absorption rate, and superior barrier performance against water vapor and gas (oxygen, hydrogen, and other gases). Its water absorption rate (<0.04%) is similar to that of glass (<0.01%). It is also chemically inert, exhibiting excellent resistance to chemicals including high temperature organic solvents and cleaning agents.

The commercial products of LCPs are various such as pure fibers, film types (Fig. 2.2), copper cladding films, and pellets. There are in several families, including brands of Vectra LCP (Celanese, Texas, U.S.A), Zenite (Celanese, Texas, U.S.A), Vecstar LCP (Kuraray, Tokyo, Japan), Vectran LCP (Kuraray, Tokyo, Japan), and copper foil-cladded Ultralam Series (Rogers Corporation, Connecticut, U.S.A). Each LCP has specific properties, for example, in Kuraray Vecstar LCP films, CT-Z is suitable for high speed transmission flexible printed circuit (FPC) and high frequency antenna because its melting temperature is 335 °C. Another Vecstar CT-F film, which is used as bonding layer in multilayer fabrication process, has relative low melting temperature 280 °C. The thickness units of commercial LCP film product are 25, 50, and 100 μm. In this study, CT-F and CT-Z films are used to make cochlear electrode array.

Table 2.1 Properties of silicon, silica, and polymeric biocompatible materials [20]

	Silicon	Glass	Polyimide	Paylene-C	PDMS	SU-8	LCP
Moisture absorption (%)	–	<0.01	0.24–4	0.06–4	0.1–1.3	0.55–0.65	<0.04
Melting temperature (°C)	1414	1252	>400	290	200–250	>300	275–335
Glass transition temperature (°C)	–	821	290–430	35–80	125–150	200–210	82–280
Young's modulus (MPa)	130,000–188,000	62750	1800–15,000	2800–3200	100–870	2000	2000–40,000
Coefficient of thermal expansion (ppm/K)	2.6	3.25	3–60	35	180–907	52	4–50
Dielectric constant	11.68 (1 kHz)	4.1 (1 MHz)	3.3 (1 MHz)	3.1 (1 kHz)	2.6 (1 MHz)	3.2 (10 MHz)	2.9 (1 MHz)

(a) **(b)**

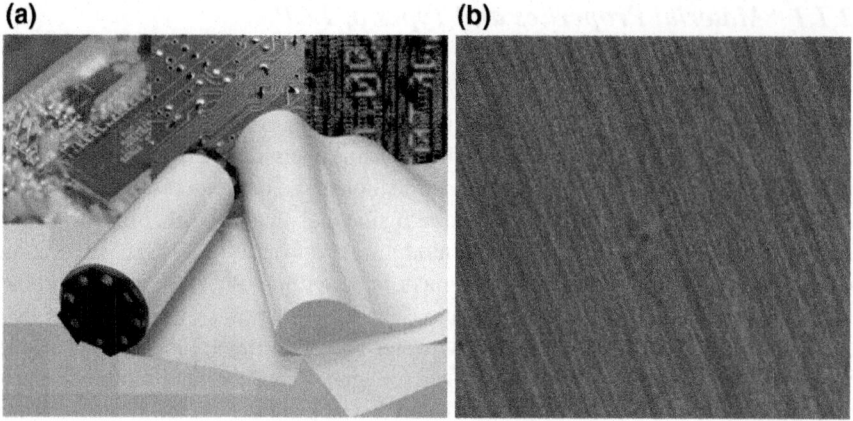

Fig. 2.2 **a** LCP film (Kuraray, Tokyo, Japan). **b** Optical image of LCP surface

Biocompatibility of LCP has been evaluated. The requirements and general testing procedures described in the International Organization for Standardization (ISO) 10993 that provides necessary guidelines for the materials to be used in medical devices have been met in some grades of Vectra LCP. Vecstar LCP-based gold microelectrode arrays met criteria of ISO 10993-5 like other polymer-based arrays [18, 21]. In vivo implantation animal test of LCP-based package revealed that the possibility of LCPs for long-term implant [17].

2.1.2 MEMS Technologies Compatible with LCP

It is highly advantageous that LCP is compatible with micro-fabrication processes using MEMS technologies such as metallization, photolithography, plasma etching, wet etching, and laser machining. Fabrication methods using MEMS technologies allow high-density microelectrode arrays and high-integration neural prosthetic systems. To use the micro-fabrication technologies, which are optimized for silicon wafer, it is necessary to attach LCP film onto silicon wafer because LCP cannot be spin-coated while polyimide and SU-8 can be. Flatness of uniform LCP surface is significant to realize ultrafine pitch of metal lines through photolithography. Spin-coated photoresist or silicone elastomer can be used to attach LCP films onto silicon wafer. Flatness is improved when using silicone elastomer by more than 50% and no bubble is found. Then, metal deposition onto LCP substrate needs surface modification technologies. Using oxygen plasma to activate LCP surface breaking C–C bond and titanium adhesion layer to form chemically strong Ti–C bond, reliable metallization onto LCP surface is possible [22]. After photolithography, wet etching, and dry etching, metal lines and electrode sites are patterned onto LCP substrate. The basic fabrication processes and scanning electron microscope (SEM) and optical

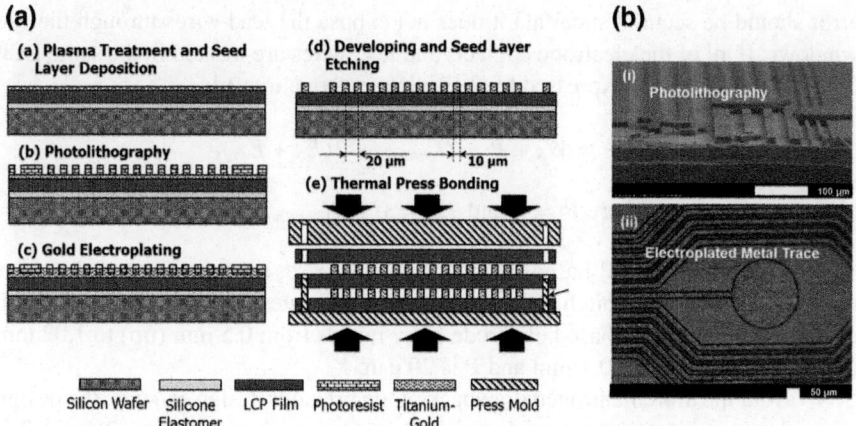

Fig. 2.3 a Basic fabrication process of LCP-based neural implants using MEMS technologies. **b** Results of photolithography (i), electroplating and seed layer etching (ii)

images of the metal patterning on LCP substrates are shown in Fig. 2.3. For insulation layer, another LCP layer can be laminated onto the patterned LCP substrate. The condition of lamination process requires enough pressure, melting temperature, and accurate lasting time. In recent years, electroplating process is added to fabricate thicker metal lines enduring lamination pressure because high-pressure results in more reliable LCP–LCP and LCP–metal bond [3]. Wavy metal lines are also used to bare high-pressure in lamination process. Through lamination process, monolithically encapsulated LCP-based devices can be achieved. Pre-cut insulation layers have been used to secure electrode site exposed for neural stimulation or recording. With the development of laser machining technology, laser ablation after lamination process has been also utilized to expose desired area and location with high-accuracy.

2.2 Cochlear Electrode Array for Atraumatic Deep Insertion

2.2.1 Electrode Design

In the LCP structure of the LCP-based cochlear electrode array, the width of the electrode (W_S), the number of channels ($N_{channel}$), the minimal line pitch (P), the error in the laser machining (electrode site: E_S, outline: E_O) and the number of LCP layers (N_{layer}) determine the total width of the electrode array (W_T). Design schemes of the previously reported LCP-based microelectrode arrays arrange stimulation sites at the center and lead wires around the stimulation sites [6]. Because the cover layer is pre-cut by the laser for site openings, the marginal spacing for lamination alignment error should be secured such that it does not expose the lead wires through the site windows. If all of the electrode contacts and lead wires are located in the same LCP film, the total width is expressed by the following equation [6]:

$$W_T = W_S + P \times N_{channel} + 2(E_S + E_O),$$

$$\text{where } P = 0 \text{ and } E_O = 0 \text{ if } N_{channel} = 1.$$

Total width of the LCP-based electrode array increases in proportion to the number of channels and the line pitch. 0.3-mm-wide electrode sites means that the total width of the 16-channel LCP-based electrode array ranges from 0.5 mm (tip) to 1.02 mm (base) if E_S and E_O are 0.1 mm and P is 20 μm.

To avoid the aforementioned drawback of the previous design scheme, the design of a side via and a cover electrode site is suggested as shown in Figs. 2.4 and 2.5. Instead of positioning an electrode at the center of the lead wires, a peripheral via so as to protect the lead wires at the center from the cutting laser beam is used. The final laser-cutting process is followed by metallization processes, sputtering and electroplating of the cover layer. Using this method, the laser-cutting of the electrode site opening windows on the cover layer only has to protect the inner lead wires because the electrode sites on the cover layer have minimal contact with the via holes. The fabrication error, $2(E_S + E_O)$, is reduced to $(E_V + E_O)$, which results in the following equation:

$$W_T = \max\left(W_S, (P \times N_{channel} + E_V + E_O)_{N_{layer}}\right),$$

$$\text{where } P = E_O = E_V = 0 \text{ if } N_{channel} = 1,$$

$$N_{channel} = N_{MWL} \text{ if } N_{layer} \geq 2,$$

$$N_{MWL} : \text{Maximum number of lead wires per layer}$$

In this case, in the current design, the width of the LCP-based electrode array with 0.3-mm-wide electrode sites ranges from 0.3 to 0.36 mm

On the basis of the minimal width described above, the LCP-based electrode array with silicone encapsulation is proposed. From channel 1 to channel 8, the contact vias and lead wires are patterned on the first substrate layer of the LCP film. The thickness of this region is 50 μm, including the inter-cover layer, except for the first channel, which is 25 μm thick because there is no cover layer. From channel 9 to channel 16, the contact vias and lead wires are patterned on the aforementioned inter-cover layer. The thickness of this basal region is 75 μm, including the cover layer.

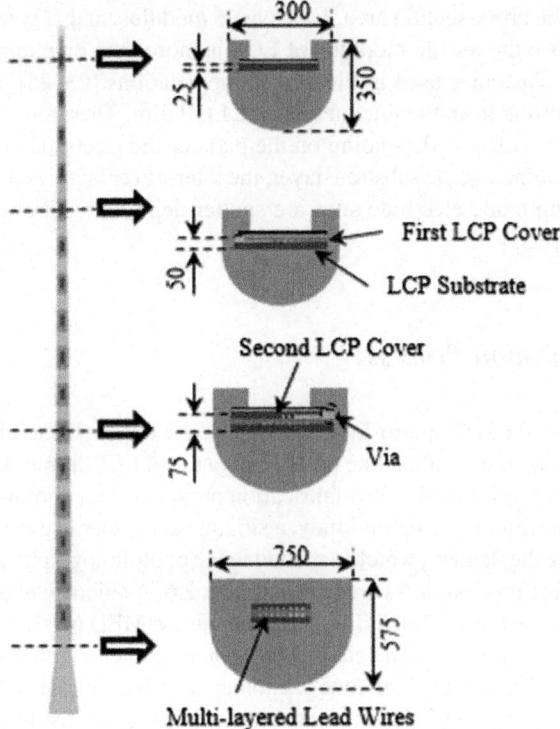

Fig. 2.4 Cross-section view of the design of tapered LCP-based cochlear electrode array for atraumatic deep insertion

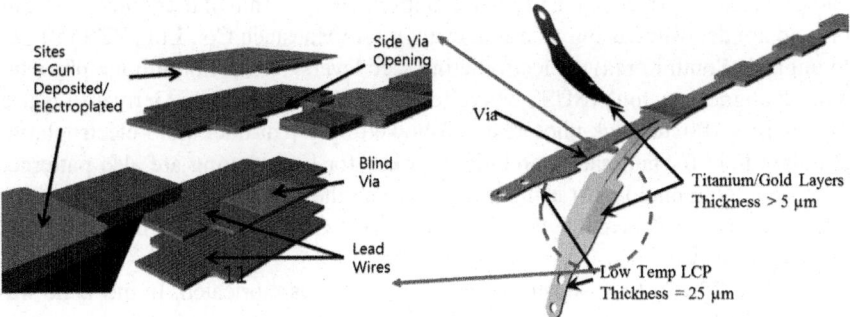

Fig. 2.5 Image of proposed cochlear electrode array using blind via

The axial stiffness (k) is expressed by the following equation

$$k = \frac{AE}{L}$$

Here, A is the cross-section area, E is tensile modulus, and L is the length of the element. Because the tensile modulus of LCP is more than one hundred times that of the silicone elastomer used in biomedical applications [23–25], the stiffness of the electrode results from the thickness of the LCP film. Therefore, it is possible to offer a variety of stiffness depending on the parts of the electrode using the current design. After laminating the substrate layer, the inter-cover layer, and the cover layer, gold and iridium oxide electrode sites are sputter-deposited and electroplated onto the exposed layers.

2.2.2   Fabrication Process

Advances in LCP-based micro-fabrication process for cochlear electrode arrays include the minimization of the line pitch using a novel LCP-film mounting method. The conventional film-based micro-fabrication process uses a photoresist to attach a film to a silicon wafer [11]. In this study, a silicone elastomer is used as an adhesion layer to enhance the flatness, which is related to the photolithography resolution [26]. Micro-fabrication process flows are shown in Fig. 2.6. A silicon wafer is spin-coated using a silicone elastomer (Nusil Silicone Technology, MED 6233, Carpinteria, CA, USA) at 2000 rpm for 70 s. The silicone elastomer is then cured on a 150 °C hot-plate for 30 min. On the cured silicone elastomer, a 25-μm-thick LCP film (Kuraray, Vecstar CT-F 25 μm, Tokyo, Japan) is attached using a roller. To clean the surface of the LCP film, the wafer onto which the LCP film is attached is immersed in acetone, methanol and isopropyl alcohol for 1 min, sequentially. Then, inductively coupled plasma (ICP) etcher for three minutes in an O_2 gaseous environment is used to clean and activate the surface of the LCP film. Subsequently, 50 nm of titanium and 150 nm of gold are deposited using an e-gun evaporator (Maestech Co., Ltd., ZZS550-2/D, Pyungtaek, South Korea) as electroplating seed layers. Next, photolithography using a mask aligner machine (SUSS MicroTec, MA6/BA6, Garching, Germany) is used to pattern a 10-μm-thick photoresist, followed by 5-μm-thick gold electroplating. An align hole for lamination and an align key for laser-cutting are also patterned. After removing photoresist using a remover and the etching of the seed layer using nitrohydrochloric acid and hydrofluoric acid, the LCP substrate is detached from the elastomer surface.

The first 25-μm LCP substrate for channels 1–8 is fabricated. In the same way, the second 25-μm LCP substrate for channels 9–16 is fabricated. Alignment holes 1 mm in diameter for the second substrate and the cover film which is the same film of the substrate are laser-cut using a UV laser machining system (Electro Scientific Industries, Flex5330, Portland, OR, USA). In addition to the align holes, the laser-cutting design should include the outlines of the via-openings, the align key for the final laser cutting process, and the contact pads. Using a thermal press machine (Carver, Model 4122, Wabash, IN, USA), the mold is heated up to 285 °C at 7 °C/min. At 285 °C, the mold maintains a pressure of 250 kg for 40 min. The LCP layers are then laminated into a united body. Therefore, as shown in Fig. 2.6f, there are different

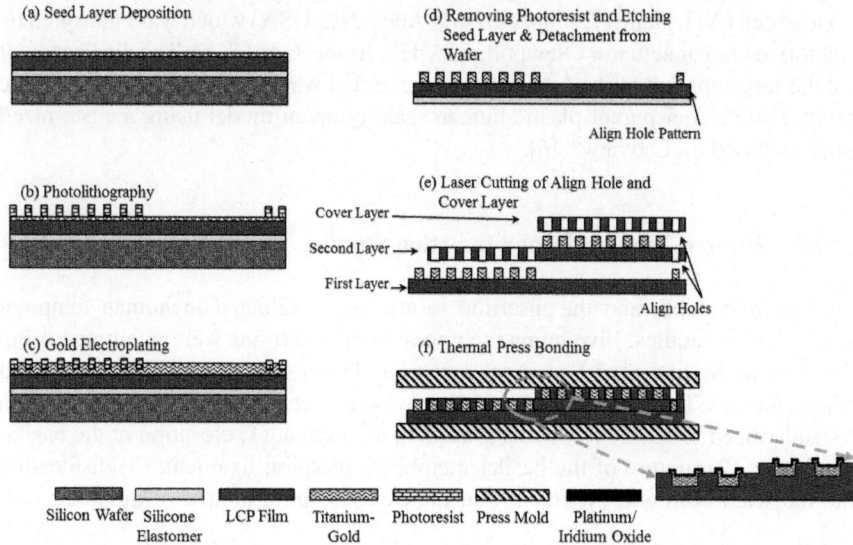

Fig. 2.6 Fabrication process flow of tapered LCP-based cochlear electrode array using MEMS technologies, thin-film process, and thermal lamination process. **a** Seed layer (Titanium and gold) deposition. **b** Photolithography using AZ 4620 photoresist. **c** Electroplating for thick metal lines. **d** Developing photoresist and wet etching seed layer using nitrohydrochloric acid and hydrofluoric acid. **e** Laser cutting of align holes and aligning LCP layers with pre-cut cover layer. **f** Thermal lamination process for monolithic encapsulation

thicknesses depending on the parts of the electrode array. The thicknesses of the LCP films are 25 μm at channel 1 (tip), 50 μm at channels 2–8 (middle), and 75 μm at channels 9 to 16 (base). A disposable shadow mask which is laser-cut using 100-μm-thick LCP film is used to deposit the site seed layer. Then, platinum or iridium oxide layers are electroplated onto the contact pads. Finally, a medical-grade silicone elastomer using a previously reported custom molding process [6] encapsulates the current LCP-based electrode array.

2.2.3 Experimental Setup and Protocol of in Vitro and in Vivo Evaluation Tests

2.2.3.1 Insertion and Extraction Force Measurements in Scala Tympani Model

The insertion force and extraction force of the finalized cochlear electrode array were measured and compared to that of a previously reported wire-based cochlear electrode array fabricated using conventional process and LCP-based cochlear electrode array which had a uniform thickness of its LCP structure. Test electrode arrays were fixed to a load cell (ATI, Nano17, Pinnacle Park, Apex, NC, USA) which was controlled by a motorized linear actuator (Newport, LTA-HS, Irvine, CA, USA). The displacement and the force applied to the load cell were recorded when the electrode arrays were inserted into a transparent plastic human scala tympani model using a customized program based on Labview® [6].

2.2.3.2   Human Temporal Bone Insertion Studies

The insertion depth and the insertion safety were evaluated in human temporal bone insertion studies. Five human temporal bone insertions were attempted using a cochleostomy and round window approach. The insertion depth was measured using a micro-CT (computerized tomography) scanned image. The insertion trauma is standardized using the following grades: 0, no trauma; (1) elevation of the basilar membrane; (2) rupture of the basilar membrane or spiral ligament; (3) dislocation into the scala vestibuli; (4) fracture of the osseous spiral lamina or modiolar wall [27].

Human temporal bone with LCP-based cochlear electrode array can be reconstructed using 3-dimensional reconstruct program (3D Slicer version 4, http://www.slicer.org) [28]. CT images are integrated in the program and the threshold values with each materials are designated. Cochlear electrode array is distinguishable in the reconstructed image (Fig. 2.7).

2.2.3.3   In Vivo Animal Study

Acute Implantation and Electrically Evoked Auditory Brainstem Response (EABR) Recording

Acute implantation of the cochlear electrode array to record ABR and EABR levels was performed on an anaesthetized guinea pig (female, 327 g) using the smartEP USB Box system (Intelligent Hearing Systems, Miami, FL, USA). The procedures used to record the ABR and EABR using this system are similar to those described in the literature [29]. To verify the hearing ability of the ear and to measure the ABR threshold, the ABR was recorded initially with sound stimuli (24 kHz tone bursts; burst amplitude: 30–70 dB SPL; burst rate: 19 Hz; burst duration: 1.6 ms). The ABR threshold was determined as the minimum sound intensity level required to evoke identifiable waves. Then, the LCP-based electrode array was inserted at a displacement of 5 mm from a round window and was fixed using biocompatible glue. The Impedance of the stimulation electrode site of the electroplated gold was 3.7 kΩ at 1 kHz. As the sound stimuli (4 kHz tone burst; burst amplitude: 100 dB SPL (Sound pressure level); burst frequency: 5 Hz; burst duration: 0.8 ms) was converted into cathodic first biphasic current stimuli (amplitude: 0.5 mA; duration: 30 μs/phase; rate: 1000/s) using an application specific integrated circuit (ASIC), the EABR was recorded using the same process used for the ABR with the cochlear

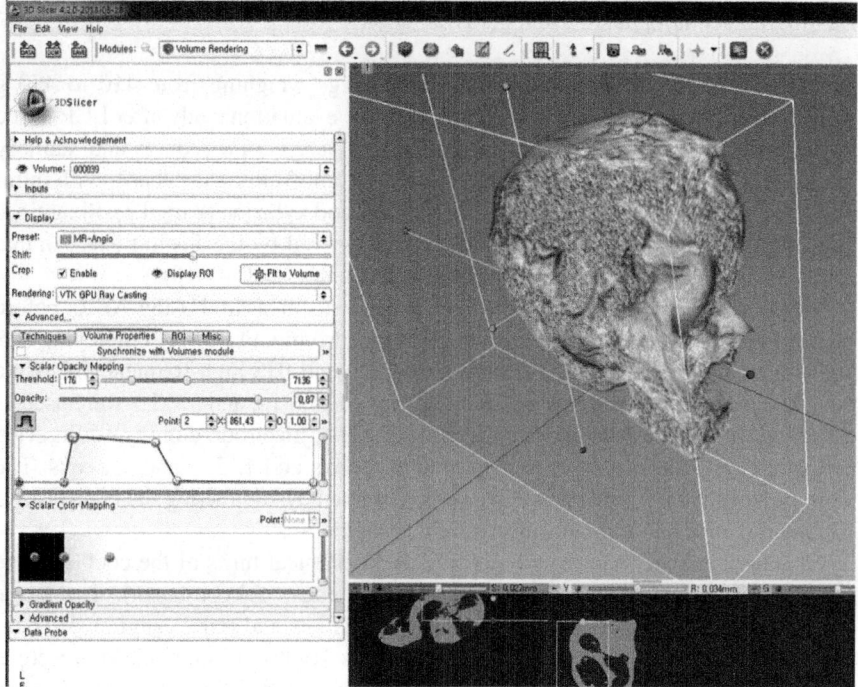

Fig. 2.7   3D slicer version 4

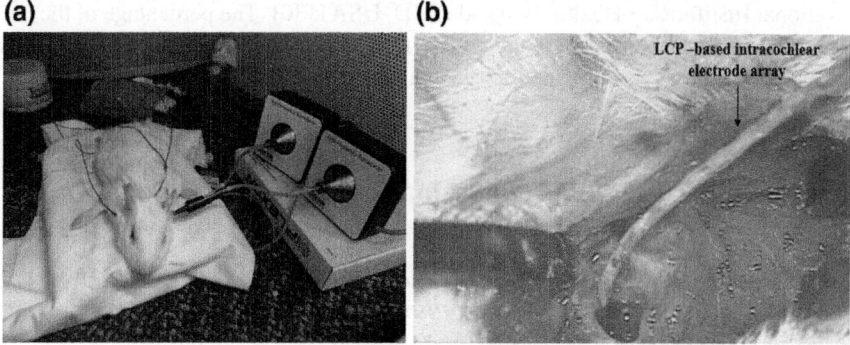

Fig. 2.8   Acute implantation of the LCP-based cochlear electrode array into guinea pig. **a** ABR recording in the soundproof room. **b** Implantation surgery

stimulator condition, i.e., on or off. All procedures conformed to protocols approved by the Institutional Animal Care and Use Committee of Seoul National University (Fig. 2.8).

Hearing Preservation and Histologic Evaluation

Seven-week-old male Dunkin-Hartley guinea pigs weighing from 410 to 450 g were used for hearing preservation and histologic evaluation study after LCP-based cochlear electrode array insertion (Fig. 2.9). All surgical procedures and auditory brainstem response (ABR) recordings were performed after anaesthesia with intra-muscular administration of ketamine (40 mg/kg) and xylazine (4 mg/kg). Six guinea pigs were randomly chosen. This study was approved by the Institutional Animal Care and Use Committee of Boramae Medical Center (2015-0010).

The hearing level of both ears of all animals was measured using ABRs before surgery. The Intelligent Hearing System (IHS Inc., Miami, FL, USA), employing HIS high-frequency transducers (HFT9911–20–0035) and IHS high-frequency software (ver. 2.33), was used to record ABRs. Tone pips of 2, 8, 16, and 32 kHz were used as the sound stimuli (5-ms duration, cos shaping, 21 Hz). Electrode-dummy was inserted and then, ABR thresholds were measured at 4 days and 1, 2, 3 and 4 weeks. The definition of the threshold shift was a value obtained by subtracting the preoperative threshold from each of the postoperative thresholds.

Histological analyses of the basal, middle and apical turns of the cochlea were performed using light microscopy (CX31; Olympus, Tokyo, Japan). The tissue responses, including intracochlear fibrosis, ossification, and spiral ganglion cell den-sity were evaluated using the results of Masson's Trichrome staining. Microscopic images of five sequential sections around the area with the most apparent tissue responses were taken and converted to JPEG files. The area of the scala tympani, and of the tissue responses surrounding it, were measured using ImageJ software (National Institutes of Health, Bethesda, MD, USA) [30]. The percentage of the area occupied by tissue responses in the scala tympani was calculated in each section and a mean percentage was obtained for each animal.

2.3   Polymer Electrode Array for Tripolar Stimulation

2.3.1   Modeling and Simulation of Polymer-Based Cochlear Electrode Array for Tripolar Stimulation

Current steering and focusing stimulation method have been studied to reduce current spread and increase distinct stimulation channel. It is important that channel interaction between stimulation channels hinders high-density electrode array and high-performance cochlear implant. Electronics can provide enormous data and information around us to internal device, but electrode array implanted in cochlea cannot accommodate and exploit the information received owing to limits of the number of stimulation channel in electrode array. Current focusing is a one of alternatives to increase effective stimulation channel without physical growth of the number of stimulation channel. Figure 2.10 explain stimulation method.

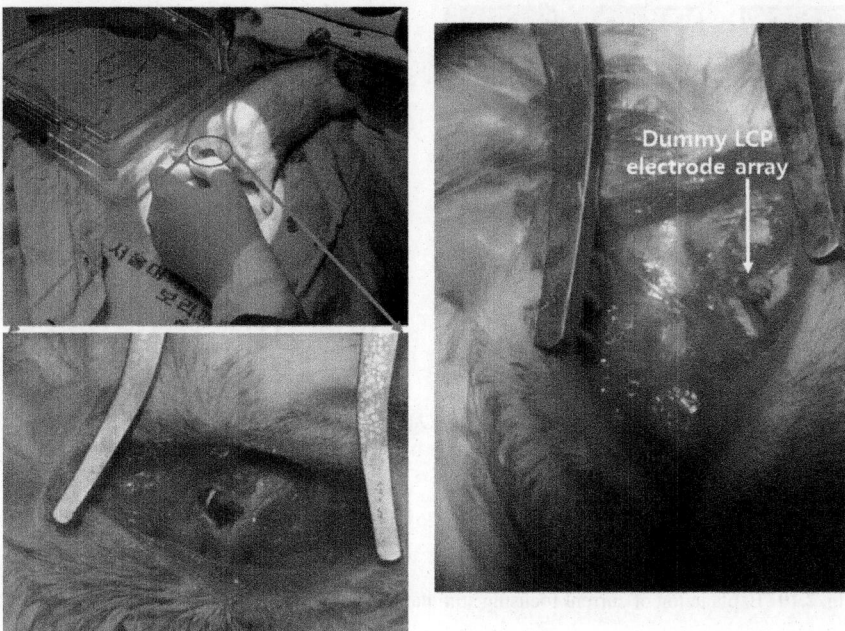

Fig. 2.9 Dummy LCP electrode array implantation surgery for hearing preservation and histological evaluation

Monopolar stimulation uses one stimulation channel and one reference channel, but multi-polymer stimulation such as bipolar and tripolar stimulation do not use the separated reference channel. Instead, multi-polar stimulation method uses fixed stimulation channel and surrounded auxiliary channel to flow current path. This flow of current prevents current spreading in the body fluid which is an electrolyte. Conventional tripolar stimulation uses 3 channels to stimulation local area. Two adjacent channels play a role of reference electrode.

More locally stimulating method through design of electrode site structure is proposed in this study. Unlike conventional electrode array, the proposed electrode array has multi-layered structure wherein the layer of auxiliary channels are separated from that of stimulation channels. This structure of electrode array is possible only in multi-layer integrated materials. Because LCP is thermoplastic material, same kind of LCP layers can be a united layer by thermo-compression lamination process.

2.3.1.1   Simulation Tool and Modeling

Commercial field element method (FEM) solver ANSYS High Frequency electromagnetic Field Stimulation (HFSS, ANSYS® Electromagnetics Suite Release 17.0.0) is used to simulate the spreading electrical field when tripolar stimulation is

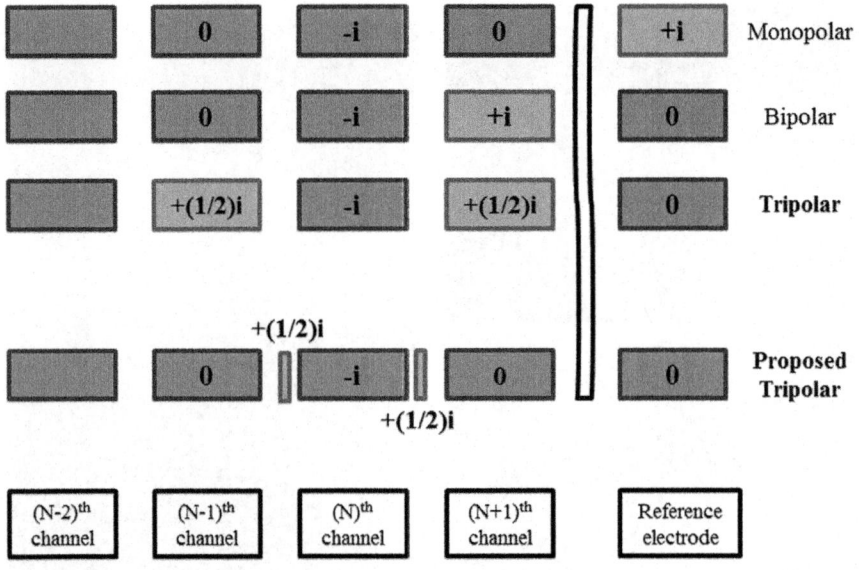

Fig. 2.10 Explanation of current focusing stimulation method

Table 2.2 Conditions of current stimulation in the simulation

Classification	Condition
Kinds of stimulation method	Monopolar, Tripolar, Locally Tripolar
Analyzing time	0–300 μs (stimulation starts at 0 s)
Phase of current pulse	Biphasic
Amplitude of current pulse	400 μA
Duration of current pulse	51 μs

implemented in the proposed electrode design. Modeling replicates human cochlea and cochlear electrode implantation. Simulation modeling of human cochlea refers to the literature [31]. The structure of scala tympani where the cochlear electrode array implanted is assumed as cylinder. Its radius is 1 mm. Perilymph, which is an extracellular fluid located in the scala tympani, and bony wall that surround cochlea are included in the modeling. The location of target cells, which are spiral ganglion cells, is 1.25 mm distant from the center of scala tympani. The radius of bony wall and analyzing space of cylindrical structure of cochlea are 1.15 and 2.5 mm, respectively.

Monopolar, tripolar, and locally tripolar (proposed stimulation structure) are simulated. The conditions of current stimulation are followed as shown in Table 2.2. In the plane where target cells are locate, relative absolute e-field amplitude at 1 mm from the position of the maximum e-field spot is calculated and analyzed (Fig. 2.11). More reduction ratio of the amplitude of e-field means more focused current stimulation.

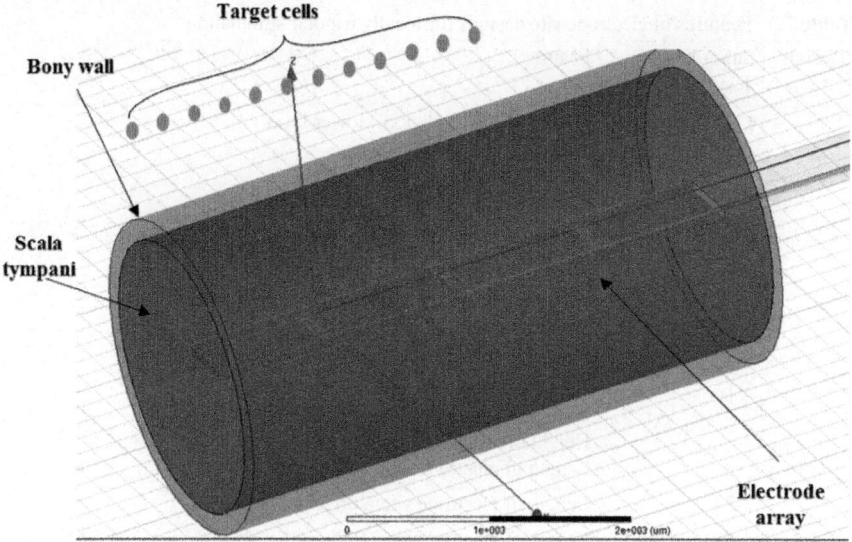

Fig. 2.11 Modeling of human cochlea and implantation of an electrode array

2.3.1.2   Electrode Designs

Electrode array is fabricated using only 25 μm-LCP films. The width and length of electrode stimulation site are 300 μm same. Only gold and LCP are used in electrode array. The thickness of electroplating (metal) is 5 μm (Center) or 10 μm (Auxiliary). Reference electrode which is needed in monopolar stimulation is 18 mm distant from the stimulation sites. Total 6 electrode site designs are simulated. After analyzing relative amplitude of e-field, one electrode design showing the most effective focused stimulation is chosen to fabricate. This design of electrode site structure is evaluated through in vitro measurement. Table 2.3 and Fig. 2.12 show the features of each electrode site designs.

The pitch of stimulation channel is 1 mm. Design 0 means general cochlear electrode array which is used in conventional monopolar and tripolar stimulation. Design 1—proposes center (stimulation) electrode site is at the lower LCP layer and auxiliary electrode site for locally tripolar stimulation is at the upper LCP layer, but Design 2 does in reverse. Design 3—suggests that center electrode site is at the lower LCP layer and auxiliary electrode site at the side wall in both sides of center electrode. There are slight differences that with and without insulation space between center site and auxiliary sites in the Design 1-1 and 1-2. Length of electrode site and distance between center site and auxiliary sites are different in the Design 3-1, 3-2, and 3-3.

Table 2.3 Features of electrode site designs for locally tripolar stimulation

Design	Subordinate design	Feature
1	1-1	Center site (stimulation): lower layer Auxiliary site: upper layer Width of electrode site: 300 μm Length of electrode site: 100 μm
	1-2	10 μm-thick insulation area between center and auxiliary site
2		Center site: upper layer Auxiliary site: lower layer Width of electrode site: 300 μm Length of electrode site: 100 μm
3	3-1	Center site (stimulation): lower layer Auxiliary site: upper layer (Side wall) Height of auxiliary site: 10 μm Length of electrode site: 100 μm
	3-2	Length of electrode site: 300 μm
	3-3	Length of electrode site: 100 μm Distance between center and side site: 100 μm

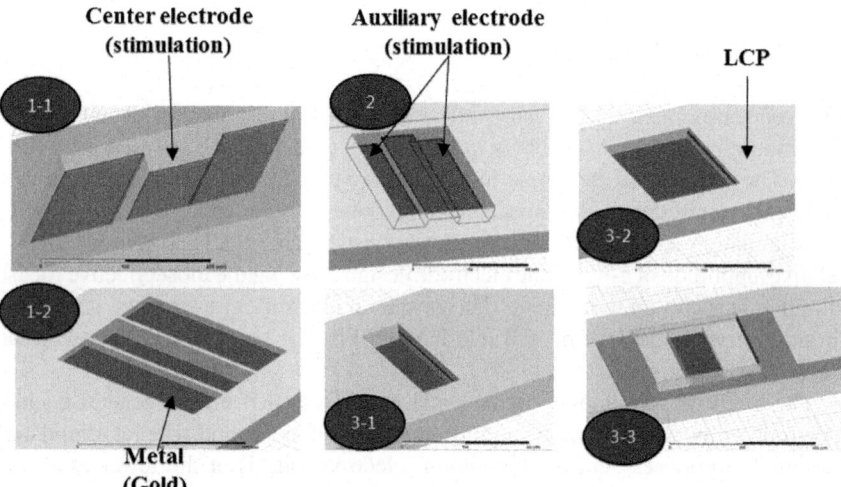

Fig. 2.12 Designs of electrode sites in the simulation

2.3.2 Fabrication Process

Fabrication process of multi-layered electrode array is more complicated than that of mono-layered electrode array. Alignment in lamination process and laser ablation process play key role in the multi-layered LCP-based electrode array for locally tripolar stimulation. To reduce width of the electrode array and enable the proposed design, via structure is necessary using laser ablation and electrical contact with conductive paste. Micromachining laser equipment (Samurai UV Marking System, DPSS Lasers Inc., Santa Clara, CA, USA) is used to make via structure and electrode site opening. Detailed fabrication process of the chosen electrode design is shown in *Result* Sect. 3.2.2.

2.3.3 In Vitro Measurements

2.3.3.1 Test Board for Tripolar Stimulation

Neural stimulator using ASICs to increase the effective number of stimulation channels via pseudo channel stimulation is implemented. The designed chip was fabricated using 0.18 μm BCDMOS process [32]. ASICs have their own 4-bit address and the data receiver converts amplitude modulated data into clock. Logical data using pulse width modulation (PWM) control ASICs to change stimulation mode (Monopolar, tripolar). The setup mode determines duration and current pulse polarity, phasic type, and current pulse amplitude. The current digital to analog converter (DAC) with 3-bit binary weighted transistors generates current in two pairs between center electrode site and auxiliary electrode sits. Ratio of the left to right pulse generator is ranging from 0:7 to 7:0 in 8 steps. The ratio used in this in vitro experiment is 3:4. As shown in Fig. 2.13, 4 ASIC chips can be loaded in the test board.

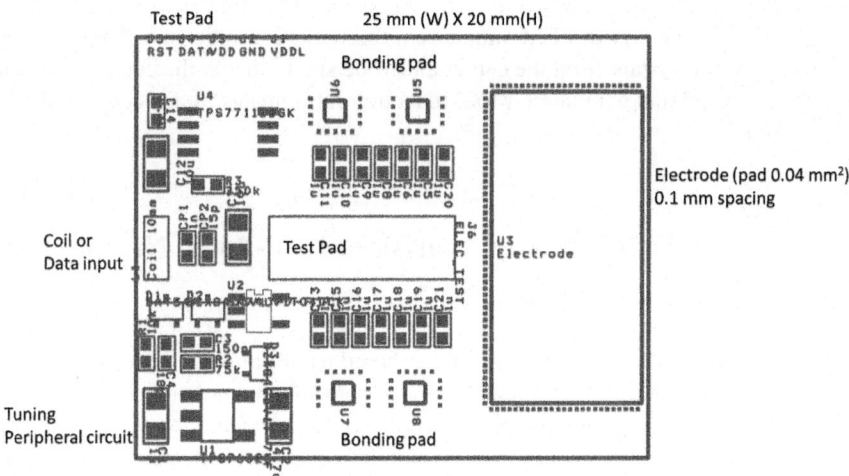

Fig. 2.13 Outline of the test board used in vitro measurements of current focusing

(a)   **(b)**

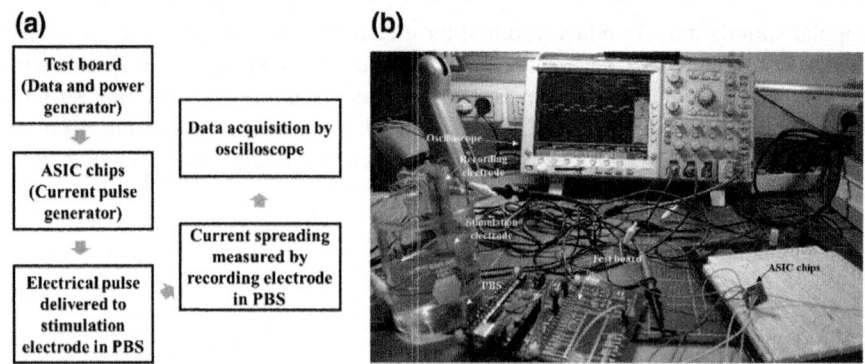

Fig. 2.14 **a** Block diagram and **b** Image of the experimental setup for in vitro current focusing measurements

2.3.3.2   Experimental Setup and Protocol

Figure 2.14 shows experimental setup for measuring current spreading in phosphate buffered saline (PBS) solution to copy body fluid. Stimulation electrode designed from simulation results is fabricated using LCP films. It is attached on a recording electrode based on printed circuit board (PCB). Test board generates data and power for ASCI chip that generates current pulse according to PWM signals. Current pulse of monopolar or tripolar stimulation mode is transferred to stimulation electrode array designed for locally tripolar stimulation. Current spreading by stimulation current pulse is recorded in the recording electrode and the measurements is saved in an oscilloscope. The space between the recording and stimulation electrode is 0.5 mm. Like analyzing method in FEM simulation, the reduction ratio of voltage level at the site of 1 mm distant from the center electrode site to that at the center electrode site is calculated and compared. More reduction ratio means more focused current stimulation.

2.4   Long-Term Reliability Analysis of LCP-Based Neural Implants

Polymer-based and hybrid metal- or silicon-based neural implant devices have been widely used to overcome the limitations of the conventional neural implant systems which are bulky, heavy, and stiff. However, these devices cannot be completely free from long-term reliability issue, because polymers are basically less hermetic than metallic and ceramic materials. Moreover, heterogeneous interfaces where water penetration occurs exist in the polymer-based neural implants. Although LCP is one of the most reliable polymers in body fluid due to its extremely low water absorption rate (<0.04%) and compatibility with monolithic seamless encapsulation process,

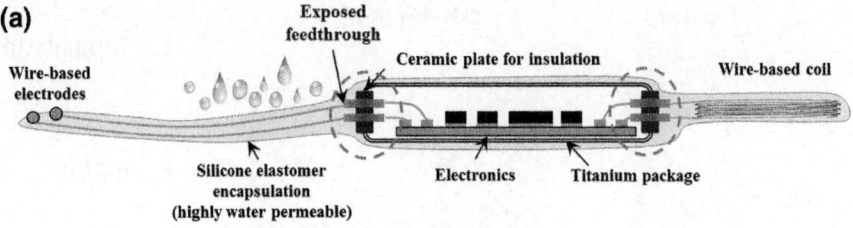

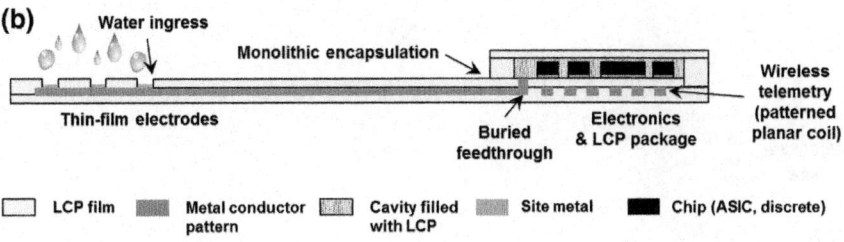

Fig. 2.15 Water penetration in **a** Metal-based neural prosthesis and **b** LCP-based neural prosthesis

it is essential for LCP-based implantable devices to evaluate and verify long-term reliability for the purpose of clinical use. In this dissertation, long-term reliability of LCP-based implantable devices are studied and analyzed synthetically from failure mechanism and technical strategies to improve reliability of LCP-based implantable devices to review of device reliability.

As shown in Fig. 2.15, LCP-based neural implant with monolithic integration of electronics package and electrode array has buried feedthrough that can be encapsulated by LCP homogeneous adhesive layers. Duration of water penetration through 0.1 mm-thick LCP surface when the relative humidity reaches 63% is calculated 64.3 years using parameters of LCP about solubility, diffusion coefficient, and moisture permeability [17]. When considering reliability of LCP-based device, water penetration into electrode array wherein LCP-LCP and Metal-LCP interface exist is more critical than that through electronics package. In this dissertation, technical strategies to improve reliability of LCP-based device is focused in electrode array.

2.4.1 Overview of the Long-Term Reliability

2.4.1.1 Failure Mechanism

In the LCP-based neural electrodes, there are three pathways through which water ingress occurs: through the LCP surface, through the interface between the LCP layers, and through the interface between the LCP and the metal (Fig. 2.16). Among these

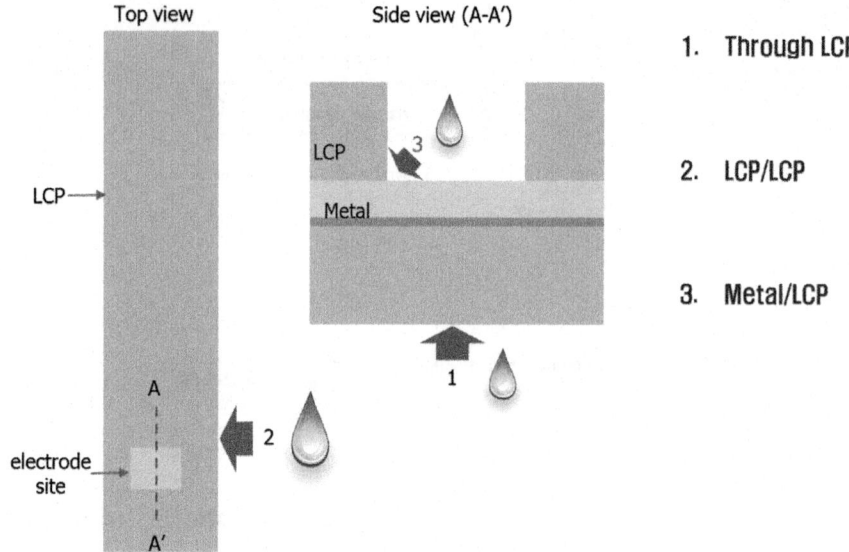

Fig. 2.16 Three pathways through which water ingress occurs

pathways, the polymer-metal interface is the limiting factor of the long-term reliability of LCP-based neural implantable devices [17]. Fundamentally, the delamination of a polymer from a metal in an electrolyte has been investigated, and studies have proven the basic delamination mechanism of polymers coated with steel [33–35]. An examination of images taken after water leakage into the electrodes verifies that the LCP-metal interface is most vulnerable to water penetration. The water penetration distance through the LCP-LCP interface is shorter than that through the LCP-metal interface.

2.4.1.2   Measurement Methods for Reliability Analysis

Helium leak measurements are the general standard method used in the industrial field to ensure the hermeticity of metal-based implantable devices. The detected helium leakage can be estimated with that of water using the square root of their ratio of the molecular mass [36] and measured leak rate can be used to calculate device lifetime using following equation.

$$t = -\frac{V}{L_{H_2O}}\left[\ln\left(1 - \frac{Q_{H_2O}}{\Delta pi_{H_2O}}\right)\right]$$

Q_{water} = the water that has leaked in the device in atm
V = the available internal volume of the package (volume of the parts inside the package should be subtracted) in cc

t = the time in seconds

L_{water} = the true water leak rate = $0.471 * L_{He}$ in atm-cc/sec

Δpi_{water} = the initial difference in the water partial pressure on the outside less the partial pressure on the inside the package (water vapor partial pressure in human body is 0.061 atm).

However, it is difficult to measure correct helium leakage in polymer-based devices accurately, as organic polymers such as LCPs naturally absorb helium [37, 38]. Besides, the range of precision of current helium leak test device is 1×10^{-13} atm-cc/sec, but, measurement level about 1×10^{-15} atm-cc/sec is needed to guarantee that LCP-based package with 0.005 cc free volume endure in relative humidity 100% environment.

Also, in terms of the electrode, water ingress from the interface between the polymer and the metal at the electrode site cannot be separated from that at the surface of the electrode. Therefore, the accelerated soak test in this study is conducted to measure and compare the long-term reliability of the LCP-based neural devices. Lifetime estimation is calculated and analyzed using Arrhenius equation and activation coefficient.

2.4.2 Technical Strategies to Improve Reliability of LCP-Based Implantable Device

2.4.2.1 Mechanical Interlocking to Strengthen Metal-LCP Adhesion

Adhesion between polymers and metals can be strengthened both mechanically and structurally. To enhance adhesion strength at the metal-polymer interface, mechanical interlocking using micro-patterning at the polymer-metal interface can be effective way [39–42]. Depending on the surface topography, the adhesion strength can be modified due to energy expenditure when a fracture occurs [39]. The proposed concept of mechanical interlocking at metal-polymer interface in the polymer-based neural electrode array is shown in Fig. 2.17. The polymer neural electrodes fabricated using consist of three parts: (1) the polymer substrate; (2) the patterned metal lead wires and electrode sites, and (3) insulating polymer with an exposed area for electrode-tissue interaction. The exposed area, which has a heterogeneous interface consisting of a metal electrode site and a polymer insulating layer, is where this study concerns. This part is where water penetration causes metal corrosion, interconnection failures, and delamination. The noble metal is used as surface for the electrode sites. However, noble metals cannot form long-term stable carbide bonds in contrast to adhesion metals such as Ti, Cr, and Ni. Alternatively, mechanical approach between a noble metal and a polymer using surface modification by means of micropatterning. The surface metal is patterned to interlock with polymer. With undercut structure of metal patterning and nanoporous surface, the noble metal can be adhered to polymer strongly.

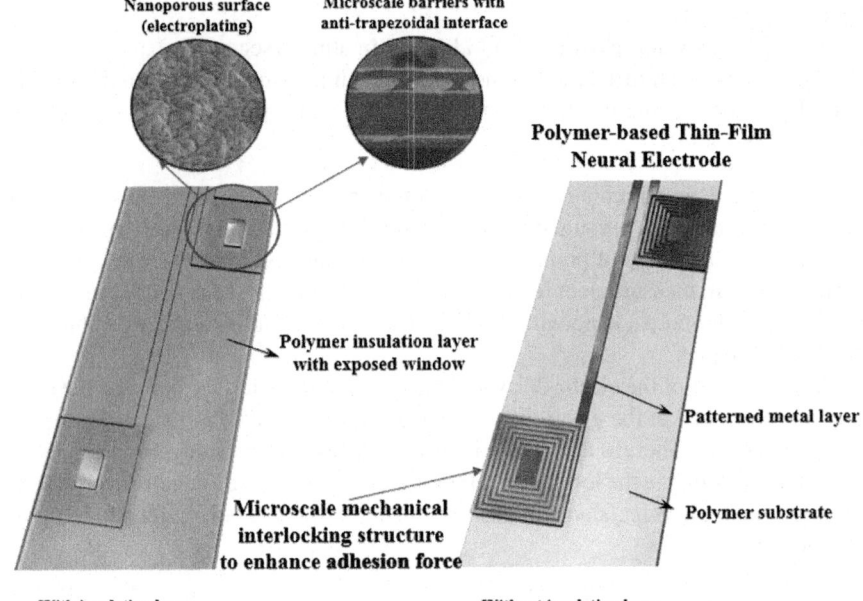

Fig. 2.17 Conceptual view of mechanical interlocking at metal-polymer interface in the polymer-based neural electrode array

Fabrication Process using Dual Lithography and Electroplating

Figure 2.18 explains the processes to fabricate the proposed LCP-based neural electrode with mechanical interlocking structure on the electrode sites. The fabrication processes using MEMS technologies include double steps of photolithography and electroplating to form the metal pattern. First, the surface of a LCP substrate (Vecstar CTF-100, Kuraray, Tokyo, Japan) attached onto a silicon wafer using a silicone elastomer (MED 6233, Nusil Silicone Technology, Carpinteria, CA, USA) is cleaned in acetone, methanol, and isopropyl alcohol, sequentially. Seed layers consisting of titanium and gold (Ti/Au) are deposited onto the surface of the LCP substrate using an e-gun evaporator (ZZS550-2/D, Maestech Co., Ltd., Pyungtaek, Korea) at 50 nm and 150 nm, respectively, after which the surface is activated by oxygen plasma using an ICP etcher for three minutes. Subsequently, a 10-μm-thick positive photoresist (Hoechst Celanese, AZ 46020, Somerville, NJ, USA) is spin-coated onto the surface of the LCP substrate (Fig. 2.18a). Then, the first photolithography step using a mask aligner machine (MA6/BA6, SUSS MicroTec, Garching, Germany) is carried out for the first electroplating of gold (Fig. 2.18b). After the removal of the remaining photoresist and the spin-coating of a new photoresist, the second photolithography step with the intentional overexposure of light to generate a trapezoidal cross-sectional photoresist and an electroplating step to create the anti-trapezoidal cross-sectional metallic structures (undercut structure) are performed using the same equipment

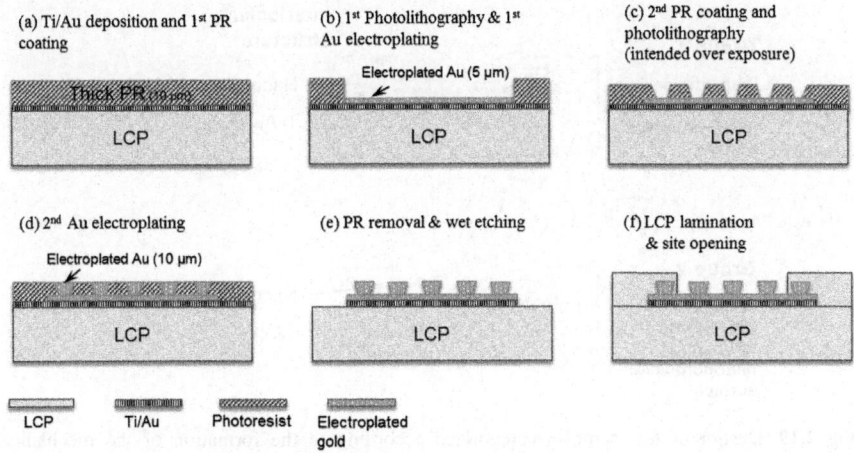

Fig. 2.18 Fabrication processes of a LCP-based neural electrode with mechanical interlocking using MEMS technologies, including oxygen plasma, e-gun evaporation, dual steps of photolithography and electroplating, and wet etching

(Fig. 2.18c, d). The remaining photoresist and the thin seed layers of Ti/Au are then removed by a wet etching process (Fig. 2.18e). Lastly, a layer of LCP film for insulation (Vecstar CTF-25, Kuraray, Tokyo, Japan) with a site window pre-cut by laser micromachining (Samurai UV Laser, DPSS Laser Inc., CA, USA) is laminated onto the patterned LCP substrate using a thermal press (Model 4122, Carver, Wabash, IN, USA) at 285 °C for 30 min while maintaining 200 kg of lamination pressure (Fig. 2.18f). During the lamination process, the LCP layers are melted into the space between the metal pattern, becoming a united body and resulting in the mechanically interlocking shapes.

Two types of LCP-based neural electrode samples are fabricated with the aforementioned methods. Group 1, with a mechanical interlocking structure, and group 2 (control group), without a microscale interlocking structure, are shown in Fig. 2.19. Both samples have electroplated nanoporous gold surface. In group 1, microscale metallic lines with a 50 μm width at 50 μm intervals are patterned by photolithography. The fabrication processes of the control group are identical to those of group 1 except for the second photolithography step to form the metal pattern for interlocking.

In Vitro Peel Test and Electrochemical Measurements

Adhesion strength is evaluated by peel test. The principle of the 180° peel test is demonstrated in earlier work [43]. The test specimens (25 × 190 × 0.2 mm) are prepared using the fabrication processes described above. The bonded area of each specimen is 25 × 70 mm². A universal testing machine (WL2100, Withlab, Daegu, Korea)

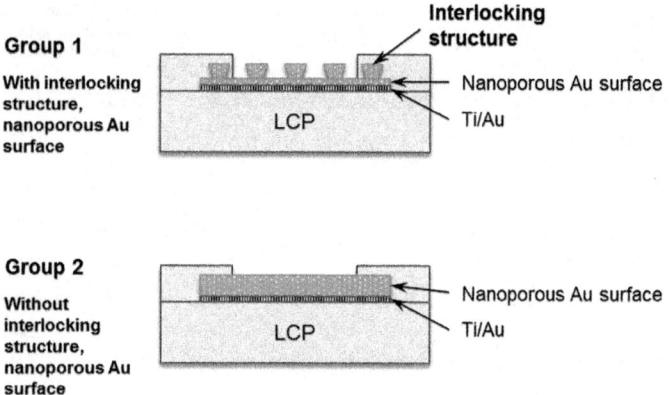

Fig. 2.19 Design of test samples categorized according to the formation of the mechanical interlocking structures

is used to peel the upper LCP film at an angle of 180° with a speed of 300 mm/min following the standard test method for the peel adhesion of pressure-sensitive tape (ASTM D3330). The peel strength (N/mm) is determined as the recorded maximum force divided by the width of the specimen. Peeling force causes the mode of failure that is classified three categories: adhesive, indicating peeling from the substrate; tearing, indicating a rupture of the upper LCP within the bonded area; or snapping, indicating that the upper LCP was ruptured away from the bonded area.

Electrochemical impedance spectroscopy (EIS) and cyclic voltammetry (CV) in PBS solutions (Invitrogen Life Technologies, Gibco 10010, Carlsbad, CA, USA) at pH 7.2 are measured using a potentiostat (Solartron Analytical, 1286 and 1287A, Farnborough, UK) and a three-cell electrochemical system with a platinum counter electrode and a silver/silver chloride (Ag/AgCl) reference electrode. EIS is measured at frequencies ranging from 10 Hz to 10 kHz using 10 mVrms of excitation. Cyclic voltammetry is also measured using an identical electrochemical cell and the electrochemical potential from −0.6 to 0.8 V versus an Ag/AgCl electrode at a rate of 100 mV/s is swept. The cathodal charge storage capacity (CSCc) is determined from the obtained cyclic voltammogram.

In Vitro Accelerated Soak Test

Accelerated soak tests are used to evaluate the long-term reliability of the implantable device. The soak tests are carried out in a 75 °C PBS condition. Although the elevated temperature for the accelerated aging of polymers is generally recommended to be at or below 60 °C [44], a temperature of 75 °C is chosen here in an effort to expedite encapsulation failure within the limited period of the experiment and to compare the "relative" reliability depending on the surface condition of the mechanical interlocking structure. The test setup for the accelerated soak test of the LCP-based neural

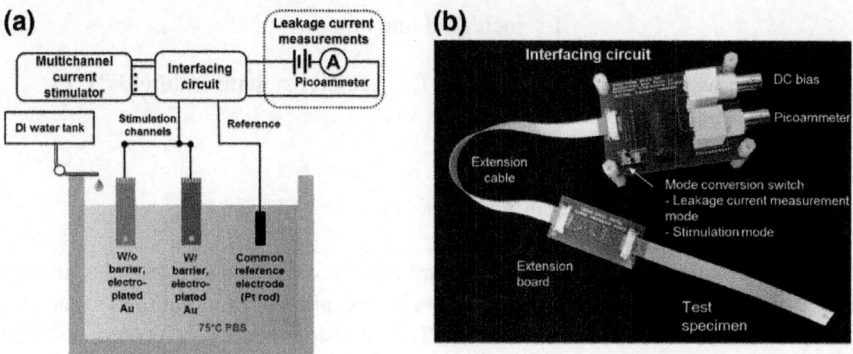

Fig. 2.20 Experimental setup for in vitro accelerated soak test. **a** Test samples are immersed in PBS at 75 °C to imitate a body environment. The test setup consists of a multichannel current stimulator to apply continuous biphasic pulses, a picoammeter for leakage current measurements, an interfacing circuit to switch between the stimulation mode and the leakage current measurement mode, and a deionized water tank to maintain the ion concentration of the PBS. **b** The interfacing circuit board, with the switch for mode conversion, and the test electrode sample are connected with an extension cable and a board

electrodes is shown in Fig. 2.20. The samples under testing are immersed with PBS-filled bottles then placed in a 75 °C convection oven. A flow regulator (DOSI-FLOW 10, Leventon, Spain) is connected to a water tank to supply deionized water to maintain the water level and the ion concentration. In order to simulate electrical stress, a customized multi-channel current stimulator generates biphasic current pulses with stimulation parameters of a 1 mA amplitude, a 32 μs duration/phase, and 1 kHz pulses per second continuously.

Device failure is determined through the presence of leakage current using inter-digitated electrodes (IDEs), with a width of 80 μm and a pitch of 100 μm, integrated into the test electrodes. Leakage currents are periodically measured in the four samples of group 1 and the five samples of group 2 using a picoammeter (Model 6485, Keithley Instruments, Inc., Cleveland, OH, USA) with 5 volts of DC bias. A customized interfacing board provides switching between the stimulation mode and the leakage current measurement mode. LabView software was used to control the picoammeter with a GPIB controller (GPIB-USB-HS, National Instruments, Austin, TX, USA). Averages and standard deviations are calculated from one hundred measurements of the leakage current value. The failure criterion for the electrode samples is set to 1 μA, which is a rough level sourced from previous work [4].

Given that the mean time to failure (MTTF) of the test samples is measured at an accelerated temperature, the lifetime at the regular body temperature (37 °C) can be calculated using the following common 'ten-degree rule', which is based on the assumption that the rate of a chemical reaction is doubled for every 10 °C elevation in the temperature.

$$f = 2^{\frac{\Delta T}{10}}, \text{ where } \Delta T = T - T_{ref}$$

f : increased rate of aging

T : elevated temperature (75 °C), T_{ref} : reference temperature(37°C)

Visual Inspection and Statistical Analysis

For visual images, optical microscopy and SEM are used to compare the junction between the LCP and the noble metal layers before and after water ingress in each test group during the accelerated soak tests. Transmission electron microscope (TEM) and focused ion beam (FIB) are also used to inspect micro-crack at the surface and interface of electrode site. All data including the peel test results, the electrochemical characterization results and the lifetime data values from the accelerated soak tests are also investigated using the Wilcoxon rank-sum test.

2.4.2.2 Fabrication using LCP and Dielectric Materials

Role of Dielectric Materials

In fabrication process of LCP-based neural implants, lamination using thermo-compression bonding press as well as photolithography is limiting factor in determining metal patterning because high-temperature and high compression pressure can affect migration of metal lines. Metal patterning can be broken if the pressure used in lamination exceeds the pressure level that metal layer can withstand. To solve this problem, electroplating to increase the thickness of metal layer is used after photolithography. However, simply increment of metal thickness has limitations in enduring pressure and pattern resolution. The line pitch resolution of metal pattern is limited to 160 μm when high-temperature and high-pressure (295 °C, 1 Mpa) is used in lamination process, which prevent high-density electrode array and interconnections. In high-temperature and high-pressure, there is another issue for migration of metal pattern after lamination process (Fig. 2.21). Low-temperature, low-pressure (285 °C, <500 kPa) lamination process can realize 20 μm-line pitch resolution, but reliability problem can arise from weak LCP-LCP interface.

  Silicon dioxide (SiO_2) on the surfaces of silicon wafer offers excellent barrier properties in flexible electronic implants [45]. Silicon nitride films deposited by plasma-enhanced CVD play a role of barrier to water and oxygen for flexible display applications [46]. Silicon neural probe have utilized triple dielectric layers to prevent water and ion penetration [47]. In this regard, fabrication process using dielectric materials with LCP is proposed to enhance reliability. Bonding strength between LCP and dielectric material is important to apply dielectric material between LCP layers. The roles of dielectric material in LCP-based implantable devices are to grip metal pattern during lamination process and delay the penetration of body fluid.

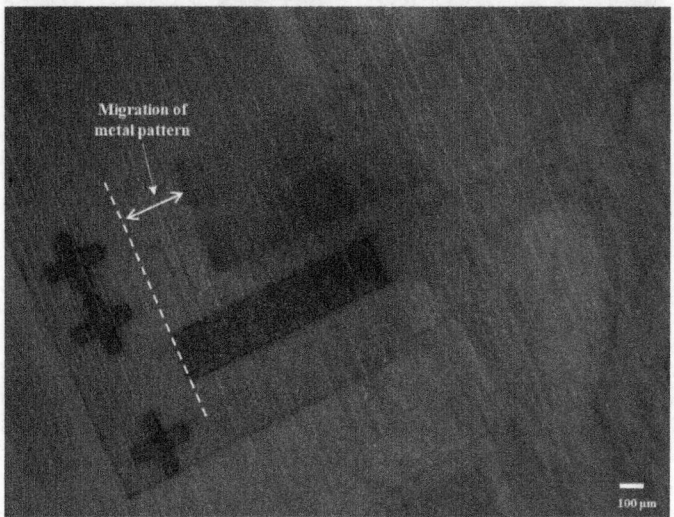

Fig. 2.21 Migration of metal pattern after lamination process

Proposed Fabrication Process

Silicon dioxide and silicon nitride are deposited on a silicon wafer using CVD process (PlasmaPro System 100, Oxford Instruments). After CVD, ICP (O_2, 100 sccm, 0.1 torr, 150 W, 3 min) is utilized to activate the surface of LCP film (Vecstar CTF-25, Kuraray) and dielectric materials for bonding enhancement. Plasma treatment increases surface hydrophilicity which affect bonding strength [48]. Identical plasma enhancement is applied in all test samples. For test samples of adhesion strength measurements, lamination process using thermo-compression press (Model 4122, Carver, Wabash, IN, USA) is carried out in low-temperature and low-pressure conditions (282 °C, <500 kPa). High-temperature and high-pressure condition is used in lamination process to evaluate that the dielectric materials prevent migration and breakage of metal lines patterned on the LCP.

Preliminary Study

Customized peel test method is utilized to compare adhesion strength between LCP and dielectric materials (Fig. 2.22). Test specimens of which width is 5 mm are fixed to a load cell (ATI, Nano17, Pinnacle Park, Apex, NC, USA) which is controlled by a motorized linear actuator (Newport, LTA-HS, Irvine, CA, USA). Peeling force is measured until the upper LCP film is peeled off 4 mm from the bottom layer and the average force is calculated during peeling off. Silicon dioxide and silicon nitride are deposited on a silicon wafer. LCP films are thermally laminated onto

Fig. 2.22 Experimental setup of customized peel test

the dielectric materials-deposited surface. Adhesion strengths of LCP-LCP (result from low-temperature, low-pressure lamination), LCP-SiO$_2$, and LCP-SiN$_x$ are compared by the means of peel test method.

Lamination process using high-temperature and high-pressure with dielectric materials is also evaluated to verify whether metal patterning is stable in dielectric material during highly reliable lamination process.

References

1. E.T. Kim, C. Kim, S.W. Lee, J.-M. Seo, H. Chung, S.J. Kim, Feasibility of Microelectrode Array (MEA) based on silicone-polyimide hybrid for retina prosthesis. Invest. Ophthalmol. Vis. Sci. **50**, 4337–4341 (2009)
2. S.W. Lee, J. Jeong, K.S. Min, S. Shin, S.B. Jun, S.J. Kim, Liquid crystal polymer (LCP), an attractive substrate for retinal implant, Sens. Mater. **24**, 2012
3. J. Jeong, S. Shin, G.J. Lee, T.M. Gwon, J.H. Park, S.J. Kim, Advancements in fabrication process of microelectrode array for a retinal prosthesis using liquid crystal polymer (LCP), in *2013 35th Annual International Conference of the IEEE Engineering in Medicine and Biology Society (EMBC)* (2013), pp. 5295–5298
4. S.W. Lee, K.S. Min, J. Jeong, J. Kim, S.J. Kim, Monolithic encapsulation of implantable neuroprosthetic devices using liquid crystal polymers, *IEEE Transactions on Biomedical Engineering*, vol. 58, 2011
5. J. Jeong, S.H. Bae, K.S. Min, J.M. Seo, H. Chung, S.J. Kim, A miniaturized, eye-conformable, and long-term reliable retinal prosthesis using monolithic fabrication of Liquid crystal polymer (LCP). IEEE Trans. Biomed. Eng. **62**, 982–989 (2015)
6. K.S. Min, S.H. Oh, M.H. Park, J. Jeong, S.J. Kim, A polymer-based multichannel cochlear electrode array. Otol. Neurotol. **35**, 1179–1186 (2014)
7. E.Y. Chow, A.L. Chlebowski, P.P. Irazoqui, A miniature-implantable RF-wireless active glaucoma intraocular pressure monitor. IEEE Trans. Biomed. Circuits Syst. **4**, 340–349 (2010)

8. T. Gwon, K. Min, J. Kim, S. Oh, H. Lee, M.-H. Park et al., Fabrication and evaluation of an improved polymer-based cochlear electrode array for atraumatic insertion. Biomed. Microdevice **17**, 1–12 (2015)

9. G.-T. Hwang, D. Im, S.E. Lee, J. Lee, M. Koo, S.Y. Park et al., In vivo silicon-based flexible radio frequency integrated circuits monolithically encapsulated with biocompatible liquid crystal polymers. ACS Nano. **7**, 4545–4553 (2013)

10. N. Laotaveerungrueng, C.H. Lin, G. McCallum, S. Rajgopal, C.P. Steiner, A.R. Rezai, et al., 3-D microfabricated electrodes for targeted deep brain stimulation, in *2009 Annual International Conference of the IEEE Engineering in Medicine and Biology Society* (2009), pp. 6493–6496

11. S.W. Lee, J.M. Seo, S. Ha, E.T. Kim, H. Chung, S.J. Kim, Development of microelectrode arrays for artificial retinal implants using liquid crystal polymers. Invest. Ophthalmol. Vis. Sci. **50**, 5859–5866 (2009)

12. C. Liu, Recent Developments in Polymer MEMS. Adv. Mater. **19**, 3783–3790 (2007)

13. P. Sattayasoonthorn, J. Suthakorn, S. Chamnanvej, J. Miao, A.G.P. Kottapalli, LCP MEMS implantable pressure sensor for intracranial pressure measurement, in *2013 IEEE 7th International Conference on Nano/Molecular Medicine and Engineering (NANOMED)* (2013), pp. 63–67

14. S.E. Lee, S.B. Jun, H.-J. Lee, J. Kim, S.W. Lee, H.-C. Shin, J.W. Chang, S.J. Kim, A flexible depth probe using liquid crystal polymer. IEEE Trans. Biomed. Eng. **59**, 2085–2094 (2012)

15. W. Xuefeng, E. Jonathan, L. Chang, Liquid crystal polymer (LCP) for MEMS: processes and applications. J. Micromech. Microeng. **13**, 628 (2003)

16. J.H. Kim, K.S. Min, S.K. An, J.S. Jeong, S.B. Jun, M.H. Cho et al., Magnetic resonance imaging compatibility of the polymer-based cochlear implant. Clin. Exp. Otorhinolaryngol. **5**, S19–S23 (2012)

17. J. Jeong, S.H. Bae, J.-M. Seo, H. Chung, S.J. Kim, Long-term evaluation of a liquid crystal polymer (LCP)-based retinal prosthesis. J. Neural Eng. **13**, 025004 (2016)

18. S.H. Bae, J.-H. Che, J.-M. Seo, J. Jeong, E.T. Kim, S.W. Lee, K.-I. Koo, G.J. Suaning, N.H. Lovell, D.-I. Cho, S.J. Kim, H. Chung, In vitro biocompatibility of various polymer-based microelectrode arrays for retinal prosthesismicroelectrode arrays for retinal prosthesis. Invest. Ophthalmol. Vis. Sci. **53**, 2653–2657 (2012)

19. T.M. Gwon, J.H. Kim, G.J. Choi, S.J. Kim, Mechanical interlocking to improve metal–polymer adhesion in polymer-based neural electrodes and its impact on device reliability. J. Mater. Sci. **51**, 6897–6912 (2016)

20. T.M. Gwon, C. Kim, S. Shin, J.H. Park, J.H. Kim, S.J. Kim, Liquid crystal polymer (LCP)-based neural prosthetic devices. Biomed. Eng. Lett. **6**, 148–163 (2016)

21. A.L. Chlebowski, Advanced radio frequency materials for packaging of implantable biomedical devices, 2009

22. J.R.N. Dean, J. Weller, M.J. Bozack, C.L. Rodekohr, B. Farrell, L. Jauniskis et al., Realization of ultra fine pitch traces on LCP substrates. IEEE Trans. Compon. Packag. Technol. **31**, 315–321 (2008)

23. C. Hassler, T. Boretius, T. Stieglitz, Polymers for neural implants. J. Polym. Sci., Part B: Polym. Phys. **49**, 18–33 (2011)

24. R. Pelrine, R. Kornbluh, J. Joseph, R. Heydt, Q. Pei, S. Chiba, High-field deformation of elastomeric dielectrics for actuators. Mater. Sci. Eng., C **11**, 89–100 (2000)

25. A.D. Woolfson, R.K. Malcolm, S.P. Gorman, D.S. Jones, A.F. Brown, S.D. McCullagh, Self-lubricating silicone elastomer biomaterials. J. Mater. Chem. **13**, 2465–2470 (2003)

26. K.S. Min, "A study on the liquid crystal polymer-based intracochlear electrode array," *Thesis Seoul National University,* 2014

27. A.A. Eshraghi, N.W. Yang, T.J. Balkany, Comparative study of cochlear damage with three perimodiolar electrode designs. The Laryngoscope **113**, 415–419 (2003)

28. A. Fedorov, R. Beichel, J. Kalpathy-Cramer, J. Finet, J-C. Fillion-Robin, S. Pujol, C. et al., Kikinis. 3D Slicer as an Image Computing Platform for the Quantitative Imaging Network. Magn. Reson. Imaging. **30**(9), 1323–1341. PMID: 22770690 (2012)

29. M. Polak, A.A. Eshraghi, O. Nehme, S. Ahsan, J. Guzman, R.E. Delgado et al., Evaluation of hearing and auditory nerve function by combining ABR, DPOAE and eABR tests into a single recording session. J. Neurosci. Methods **134**, 141–149 (2004)

30. C.A. Schneider, W.S. Rasband, K.W. Eliceiri, NIH image to imageJ: 25 years of image analysis. Nature methods **9**(7), 671–675 (2012)

31. B.H. Bonham, L.M. Litvak, Current focusing and steering: modeling, physiology, and psychophysics. Hear. Res. **242**, 141–153 (2008)

32. J.H. Park, C. Kim, S.H. Ahn, T.M. Gwon, J. Jeong, S. Beom Jun, S.J. Kim, A distributed current stimulator ASIC for high density neural stimulation, in *2016 38th Annual International Conference of the IEEE Engineering in Medicine and Biology Society (EMBC)*, 2016

33. A. Leng, H. Streckel, K. Hofmann, M. Stratmann, The delamination of polymeric coatings from steel Part 3: effect of the oxygen partial pressure on the delamination reaction and current distribution at the metal/polymer interface. Corros. Sci. **41**, 599–620 (1998)

34. A. Leng, H. Streckel, M. Stratmann, The delamination of polymeric coatings from steel. Part 2: first stage of delamination, effect of type and concentration of cations on delamination, chemical analysis of the interface. Corros. Sci. **41**, 579–597 (1998)

35. A. Leng, H. Streckel, M. Stratmann, The delamination of polymeric coatings from steel. Part 1: Calibration of the Kelvinprobe and basic delamination mechanism. Corros. Sci. **41**, 547–578 (1998)

36. A. Vanhoestenberghe, N. Donaldson, The limits of hermeticity test methods for micropackages. Artif. Organs **35**, 242–244 (2011)

37. K. Aihara, M.J. Chen, C. Cheng, A.V.H. Pham, Reliability of liquid crystal polymer air cavity packaging. IEEE Trans. Compon. Packag. Manuf. Technol. **2**, 224–230 (2012)

38. A.-V. Pham, Packaging with liquid crystal polymer. IEEE Microwave Mag. **5**, 83–91 (2011)

39. W.-S. Kim, I.-H. Yun, J.-J. Lee, H.-T. Jung, Evaluation of mechanical interlock effect on adhesion strength of polymer–metal interfaces using micro-patterned surface topography. Int. J. Adhes. Adhes. **30**, 408–417 (2010)

40. F.K. LeGoues, B.D. Silverman, P.S. Ho, The microstructure of metal–polyimide interfaces. J. Vac. Sci. Technol., A **6**, 2200–2204 (1988)

41. D.E. Packham, Surface energy, surface topography and adhesion. Int. J. Adhes. Adhes. **23**, 437–448 (2003)

42. J.D. Venables, Adhesion and durability of metal-polymer bonds. J. Mater. Sci. **19**, 2431–2453 (1984)

43. J.F. McCabe, T.E. Carrick, H. Kamohara, Adhesive bond strength and compliance for denture soft lining materials. Biomaterials **23**, 1347–1352 (2002)

44. W. Chun, N. Chou, S. Cho, S. Yang, S. Kim, Evaluation of sub-micrometer parylene C films as an insulation layer using electrochemical impedance spectroscopy. Prog. Org. Coat. **77**, 537–547 (2014)

45. E. Song, H. Fang, X. Jin, J. Zhao, C. Jiang, K.J. Yu et al., Thin, transferred layers of silicon dioxide and silicon nitride as water and ion barriers for implantable flexible electronic systems. Adv. Electron. Mater. **3**, 1700077 (2017)

46. D.S. Wuu, W.C. Lo, C.C. Chiang, H.B. Lin, L.S. Chang, R.H. Horng et al., Water and oxygen permeation of silicon nitride films prepared by plasma-enhanced chemical vapor deposition. Surf. Coat. Technol. **198**, 114–117 (2005)

47. S.J. Oh, J.K. Song, S.J. Kim, Neural interface with a silicon neural probe in the advancement of microtechnology. Biotechnol. Bioprocess Eng. **8**, 252–256 (2003)

48. A.U. Alam, Y. Qin, M.R. Howlader, M.J. Deen, Direct bonding of liquid crystal polymer to glass. RSC Adv. **6**, 107200–107207 (2016)

Chapter 3
Results

3.1 LCP-Based Cochlear Electrode Array for Atraumatic Deep Insertion

3.1.1 Fabricated Electrode Array

LCP-based cochlear electrode array for atraumatic deep insertion is fabricated using MEMS technologies and thin-film process. The number of stimulation channel is sixteen and two reference electrodes are at 72 mm apart from the end of stimulation channel. The lengths of electrode array is 28 mm and the thickness of the current LCP structure varies depending on the part of the cochlear electrode array, which is 25 μm at the tip, 50 μm at the middle, and 75 μm at the base. The diameters of the electrode array are 0.3 mm (tip) and 0.75 mm (base). The dimensions of the current LCP-based cochlear electrode array are similar or finer than that of a conventional cochlear electrode array due to peripheral vias and via-openings (Fig. 3.1).

3.1.2 Insertion and Extraction Force Measurements

The average of measured insertion force of the electrode array, shown in Fig. 3.2, is 2.4 mN at a displacement of 8 mm from the round widow. The average measured maximum extraction force is 33.0 mN as shown in Fig. 3.3.

© Springer Nature Singapore Pte Ltd. 2018
T. M. Gwon, *A Polymer Cochlear Electrode Array: Atraumatic Deep Insertion, Tripolar Stimulation, and Long-Term Reliability*, Springer Theses,
https://doi.org/10.1007/978-981-13-0472-9_3

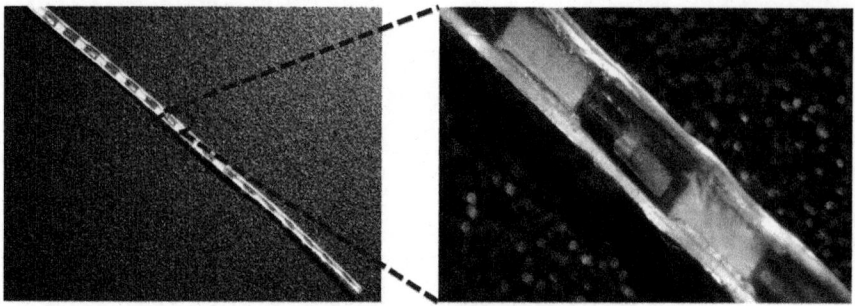

Fig. 3.1 LCP-based cochlear electrode array with tapered structure

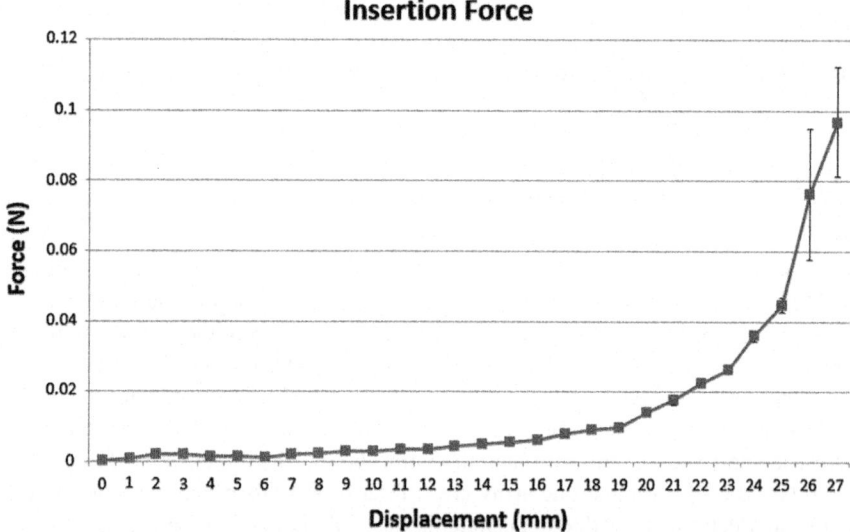

Fig. 3.2 Insertion force measured in the plastic scala tympani model

3.1.3   Insertion Trauma in Human Temporal Bone Insertion Study

In the present human temporal bone insertion studies, the insertion depths of the electrode arrays in the human temporal bone, which are measured in micro-CT images, are 450°, 450°, 360° and 630° in the round window approach and 495° in the cochleostomy approach. Insertion trauma is investigated in the cross-section view of enucleated cochlea. Five electrode arrays are inserted without trauma (grade 0) at the first turn. However, for the 630° insertion, dislocation into the scala vestibuli (grade 3) is observed at the tip of the electrode array. There is no observable trauma in the other cases (Figs. 3.4, 3.5, 3.6, 3.7 and 3.8).

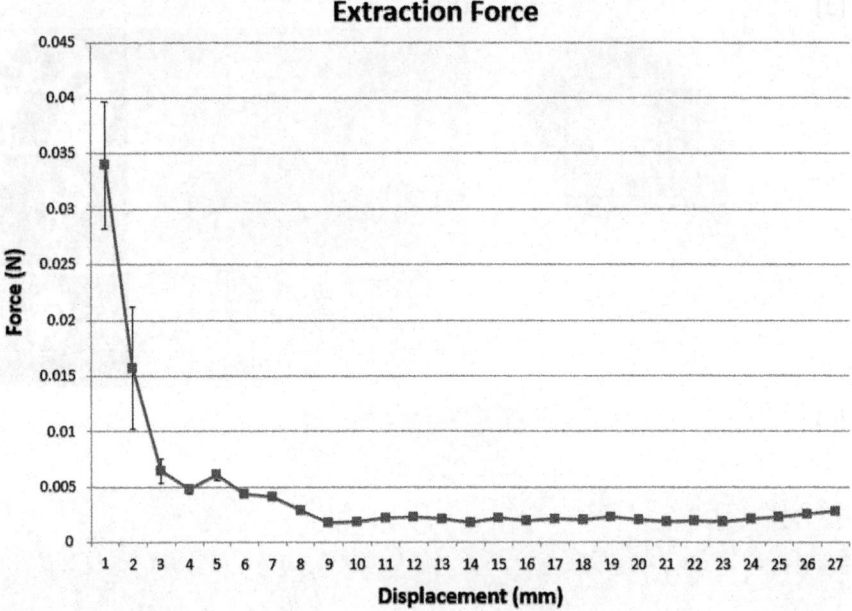

Fig. 3.3 Extraction force measured in the plastic scala tympani model

3.1.4 Electrically Evoked Auditory Brainstem Response Recording

Impedance amplitude and phase angle of the stimulation electrode site of the nanoporous electroplated gold at frequencies ranging from 0.1 to 100 kHz indicates 3.66 kΩ and $-43.2°$ at 1 kHz. A 24 kHz-tone burst-evoked ABR is shown in Fig. 3.9 wherein the ABR threshold is measured at 35 dB SPL. The EABR recordings, which are recorded when the current stimulator was on and off, respectively, are depicted in Fig. 3.10. The measured peak-to-peak amplitudes are 9.2 and 1.93 μV without filtering, respectively. When operating the current stimulator, the EABR waveform shows positive peak value that is a meaningful neural signal. When the stimulator is off, there is a significant change in the EABR waveform. The signal magnitude becomes much weaker and there is no obvious neural signal. It is assumed and evaluated that the guinea pig is not able to recognize the sound stimuli.

3.1.5 Histological Change and Hearing Preservation

There are shifts in ABR threshold at all frequency measured. Threshold shift levels at all frequency range from 25.6 to 38 dB. The mean and standard deviation of the

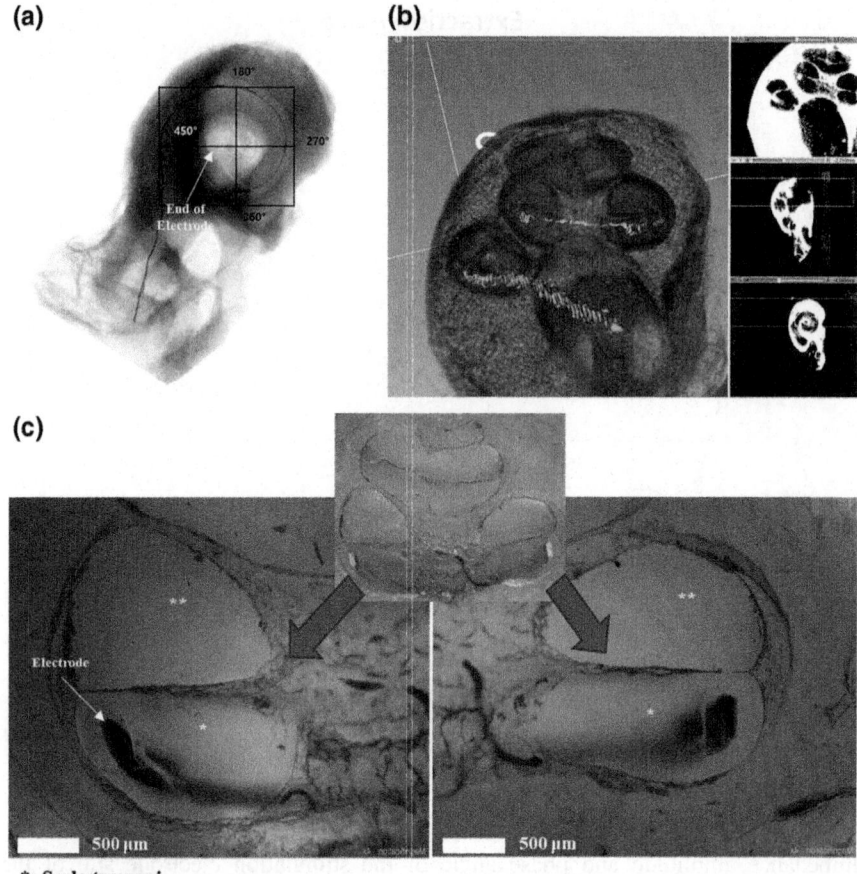

(a)

(b)

(c)

*: Scala tympani
**: Scala vestibuli

Fig. 3.4 Human temporal bone insertion study of electrode 1 (Round window approach). **a** Micro-CT image for insertion depth (450°). **b** Reconstructed 3-D image. **c** Cross-sectional views of cochlea where LCP electrode array is inserted (Trauma 0)

shift results are arranged in Fig. 3.11. The threshold shift levels have been maintained during 4 weeks.

Cochlea is enucleated after 4 weeks implantation, as shown in Fig. 3.12. In the cross-sectional view of cochlea, dummy LCP cochlear electrode array is inserted in the scala tympani without any trauma. Stable implantation has been maintained during 4 weeks.

In Fig. 3.13, there is fibrosis at the site where cochlear electrode array was inserted. Using Masson's Trichrome, stained color is different in each tissue: collagen fiber and bone to blue, Muscle fibers, cytoplasm, erythrocytes, keratin, and fibrin to red, and nuclei to black or blue. White circle areas indicate fibrosis.

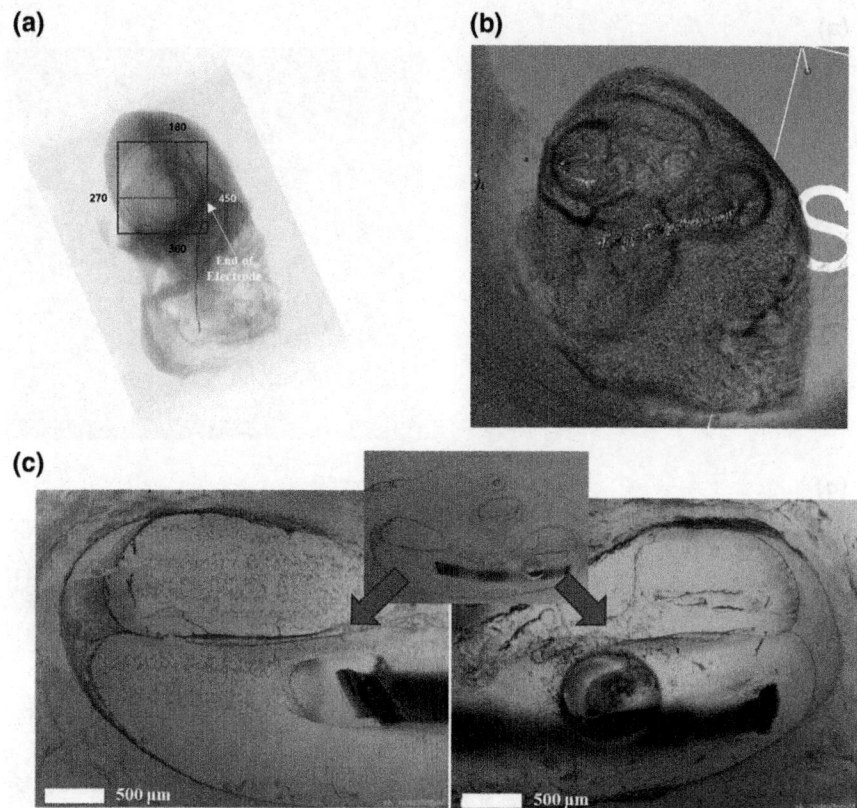

Fig. 3.5 Human temporal bone insertion study of electrode 2 (Round window approach). **a** Micro-CT image for insertion depth (450°). **b** Reconstructed 3-D image. **c** Cross-sectional views (Trauma 0)

3.2 Polymer Electrode Array for Tripolar Stimulation

3.2.1 Simulation Results According to Electrode Site Design

Figure 3.14 shows the simulation results depending on electrode site designs. Center electrode site for stimulation is at the lower layer in Design 1 and the upper layer in Design 2. Design 3 utilizes the side wall electrode site for tripolar stimulation. The value of e-field at the plane where spiral ganglion cells exist are calculated using finite element method. The obtained values are arranged in the Tables 3.1 and 3.2. According to the simulation results, Design 3-3 shows the most effective focused stimulation to be chosen to fabricate. The fabricated electrode structure is evaluated in in vitro test setup.

(a)   **(b)**

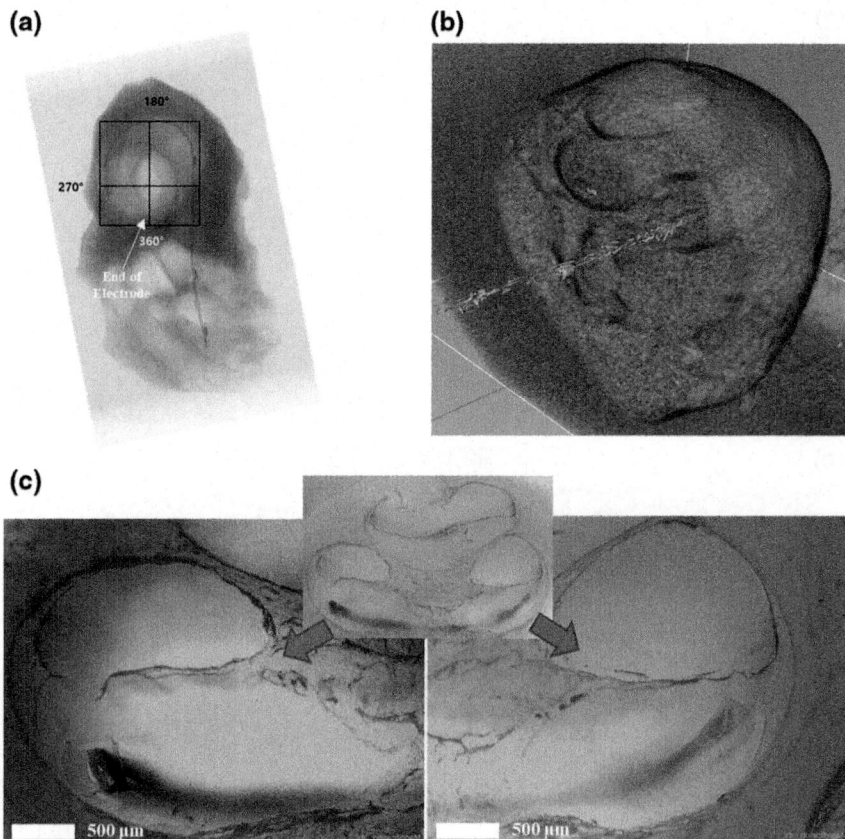

Fig. 3.6 Human temporal bone insertion study of electrode 3 (Round window approach). **a** Micro-CT image for insertion depth (360°). **b** Reconstructed 3-D image. **c** Cross-sectional views (Trauma 0)

3.2.2   Fabricated Electrode Array

Electrode array following Design 3-3 can be fabricated using laser ablation technology and multi-layered through via structure. The fabrication process and result at each process are presented in Figs. 3.15 and 3.16. Via at back side of micro-patterned LCP film is made using laser ablation. Silver paste (EPO-TEK H20E, Epoxy Technology Inc., MA, USA) is selected in this fabrication process as a material for via-fill due to its easy curing process. Four-channel LCP-based electrode array for locally tripolar stimulation is fabricated. Pitch of the stimulation channel is 1 mm and the distance between center and auxiliary site is 100 μm. The width of electrode sites is 300 μm.

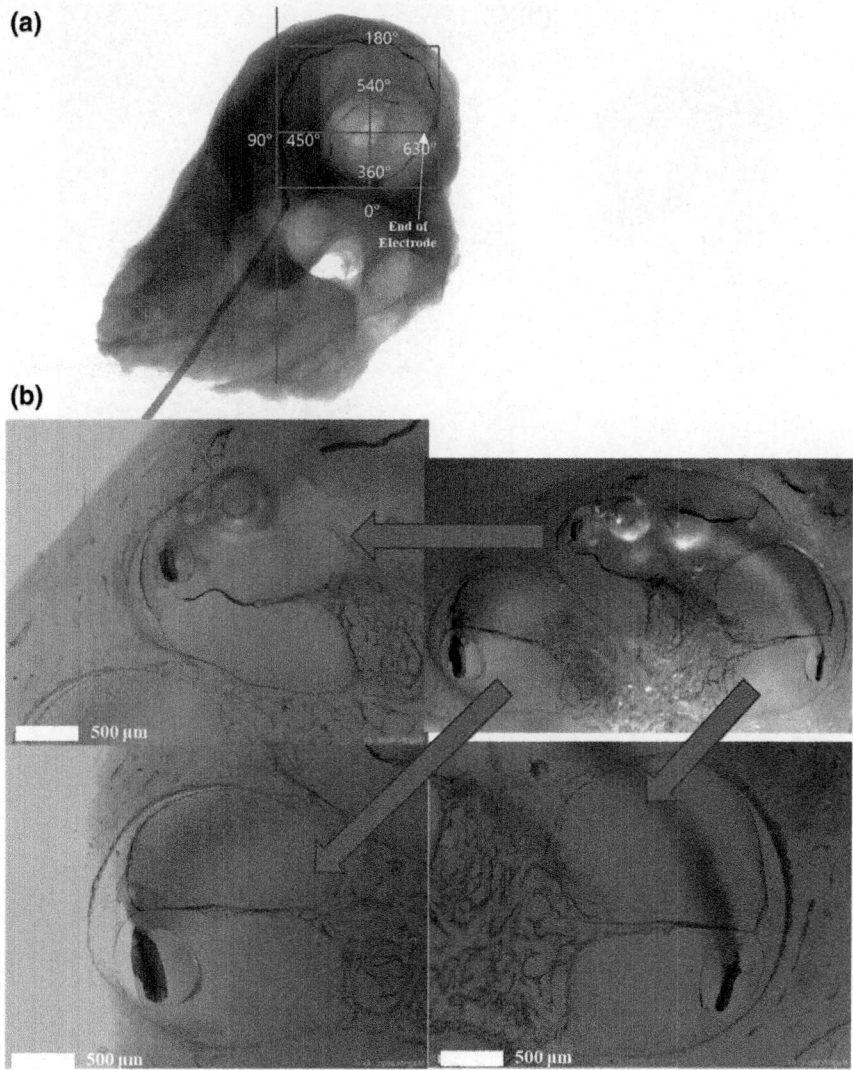

Fig. 3.7 Human temporal bone insertion study of electrode 4 (Round window approach). **a** Micro-CT image for insertion depth (630°). **b** Cross-sectional views (Trauma 3)

3.2.3 In Vitro Measurements

Biphasic current pulses are generated by ASIC chip. PWM data to generate current pulses from test board is delivered to ASIC chip. Duration of cathodic pulse of biphasic current pulse is 120 μs and amplitude of pulses is 1.24 mA in monopolar

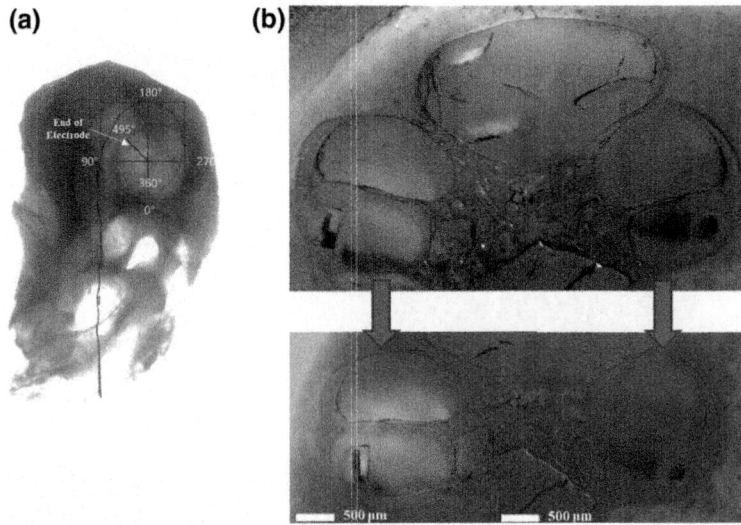

Fig. 3.8 Human temporal bone insertion study of electrode 5 (Cochleostomy approach). **a** Micro-CT image for insertion depth (495°). **b** Cross-sectional views (Trauma 0)

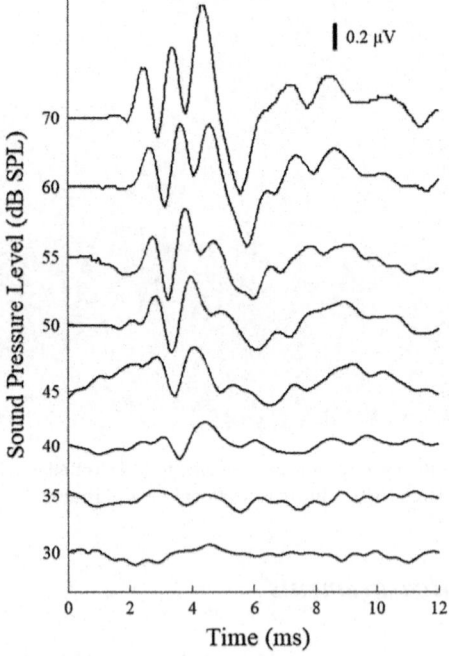

Fig. 3.9 ABR recording which indicates the threshold level at 35 dB SPL

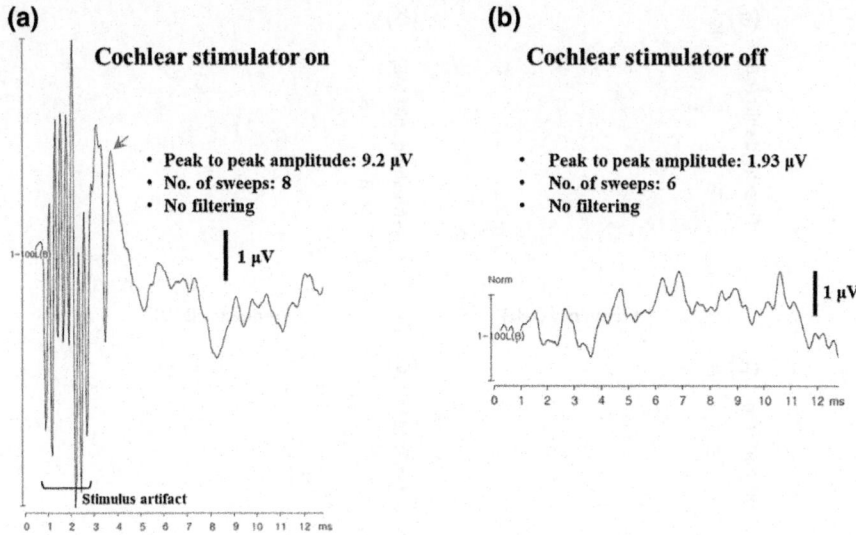

Fig. 3.10 Recorded EABR with Stimulation **a** On and **b** Off

stimulation mode. In tripolar stimulation mode, the current is divided into ratio of 5.8:4.8 at left and right auxiliary electrode, respectively (Fig. 3.17).

Reduction ratio is compared according to stimulation methods. The proposed locally tripolar stimulation using multi-layered electrode array indicates the most effective focused stimulation, which shows imbalanced spread of stimulation due to the position of reference electrode (Table 3.3).

3.3 Long-Term Device Reliability

3.3.1 LCP-Based Neural Electrode Array Using Mechanical Interlocking at Metal-LCP Interface

3.3.1.1 Fabricated Electrode Array and Metal-LCP Interface

Metal patterning for mechanical interlocking between noble metal and LCP is fabricated using double steps of photolithography and electroplating. Surface of electrode site is nanoporous electroplated gold. All test samples have IDE pattern for current detection in accelerated soak tests. The microscale metallic lines composed of both a width of 50 μm and pitch of 50 μm are successfully patterned on the electrode site using double steps of photolithography and electroplating. IDEs with a width of 80 μm and a pitch of 100 μm is patterned onto the LCP substrate to measure leakage current indicating device failures (Fig. 3.18).

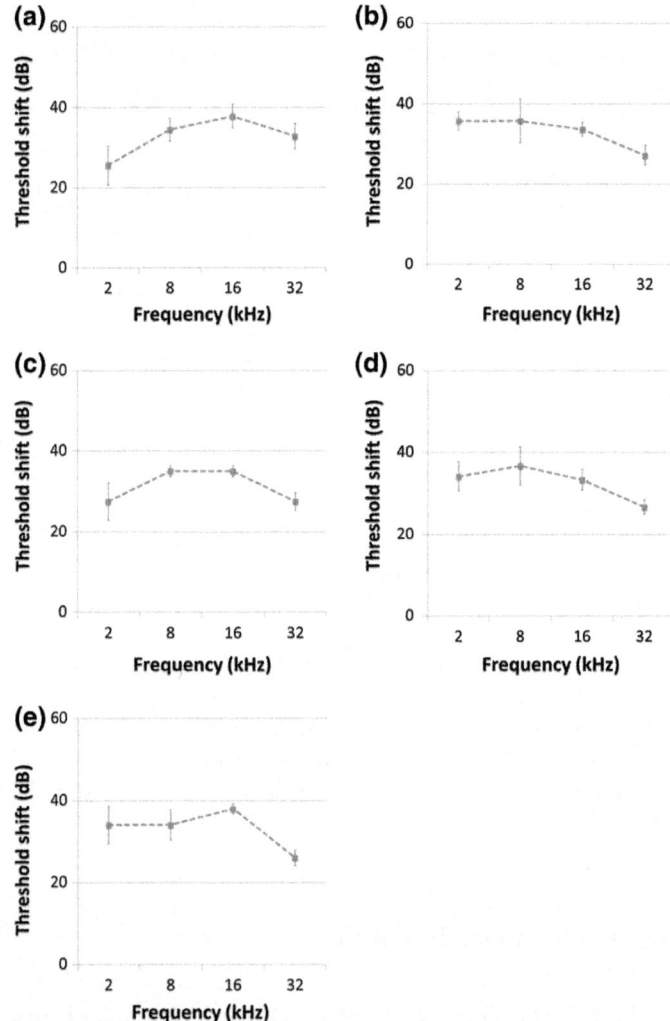

Fig. 3.11 ABR threshold shift after **a** 4 days, **b** 1 week, **c** 2 weeks, **d** 3 weeks, and **e** 4 weeks implantation

As shown in Fig. 3.19, the pre-cut LCP insulation layer with exposed area is aligned to the center of electrode site and laminated onto the LCP substrate using thermal compression bonding. It is shown in cross-section of electrode site that undercut metal patterns surrounded by LCP film are mechanically interlocked with the upper LCP insulation layer. Moreover, LCP layers form monolithic seamless adhesion and bond to the interlocking structures without voids.

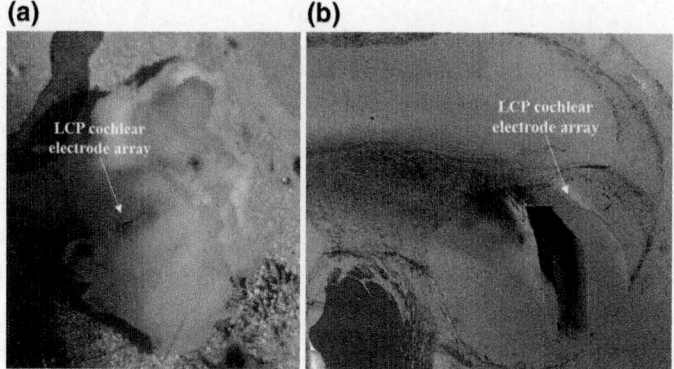

Fig. 3.12 a Enucleated cochlea after 4 weeks implantation. **b** Cross-section of the cochlea with LCP electrode implantation

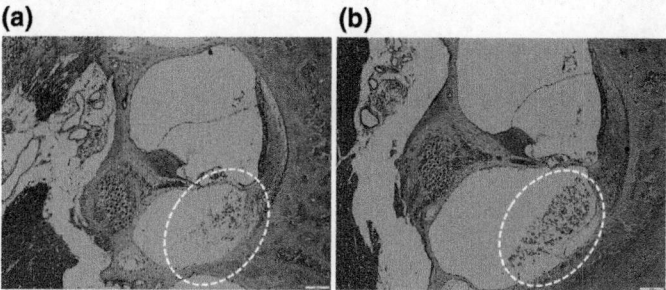

Fig. 3.13 Cross-section view of stained cochlea in **a** Middle and **b** Basal turn

3.3.1.2 Adhesion Force and Electrochemical Measurements

Table 3.4 shows results of peel test to evaluate adhesion force. There were specific differences in peel strength between the test group 1 with interlocking structures and the control group 2 ($p < 0.1$). In comparison with the control group, the maximum force per width of the test specimens with interlocking interface between LCP and noble metal has increased by 34.8%.

Figures 3.20 and 3.21 show electrochemical measurements of test samples. Group 1 (with mechanical interlocking) shows lower amplitude of impedance and larger CSCc than Group 2 ($p < 0.01$). Impedances at 1 kHz are measured as 356.99 and 474.31 Ω in group 1 and 2, respectively. The measured and calculated cathodal charge storage capacities are 1.388 mC cm^{-2} in group 1 and 1.031 mC cm^{-2} in group 2.

(a)   **(b)**

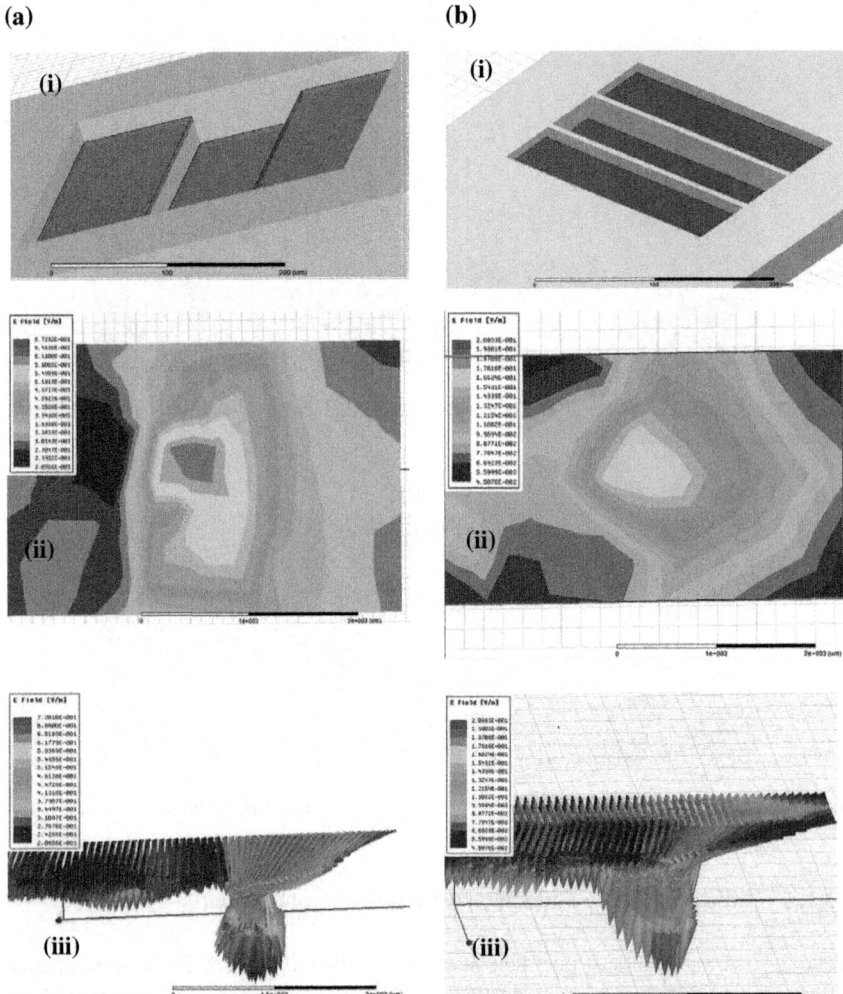

Fig. 3.14 Simulation results of tripolar stimulation according to electrode designs. **a** Design 1-1, **b** Design 1-2, **c** Design 2, **d** Design 3-1, **e** Design 3-2, **f** Design 3-3 [(i) electrode structure, (ii) absolute value of e-field at the plane of analyzing, (iii) picture of vector distribution of e-field]

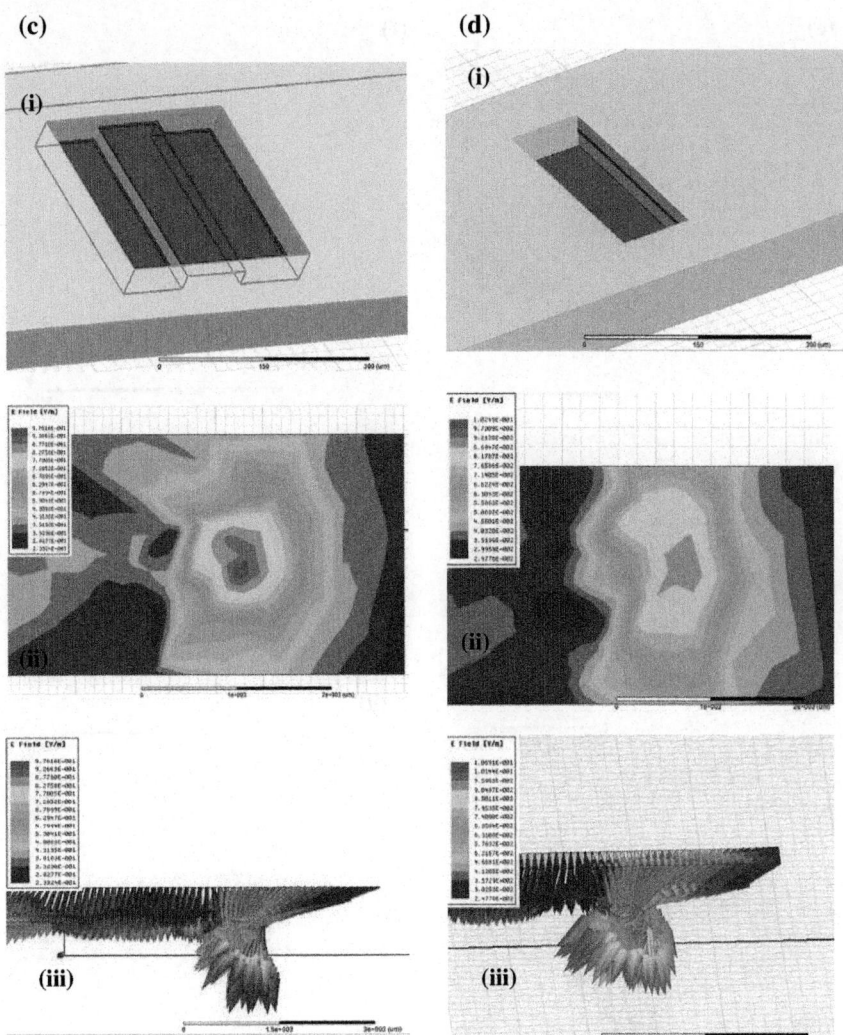

Fig. 3.14 (continued)

(e)

(f)

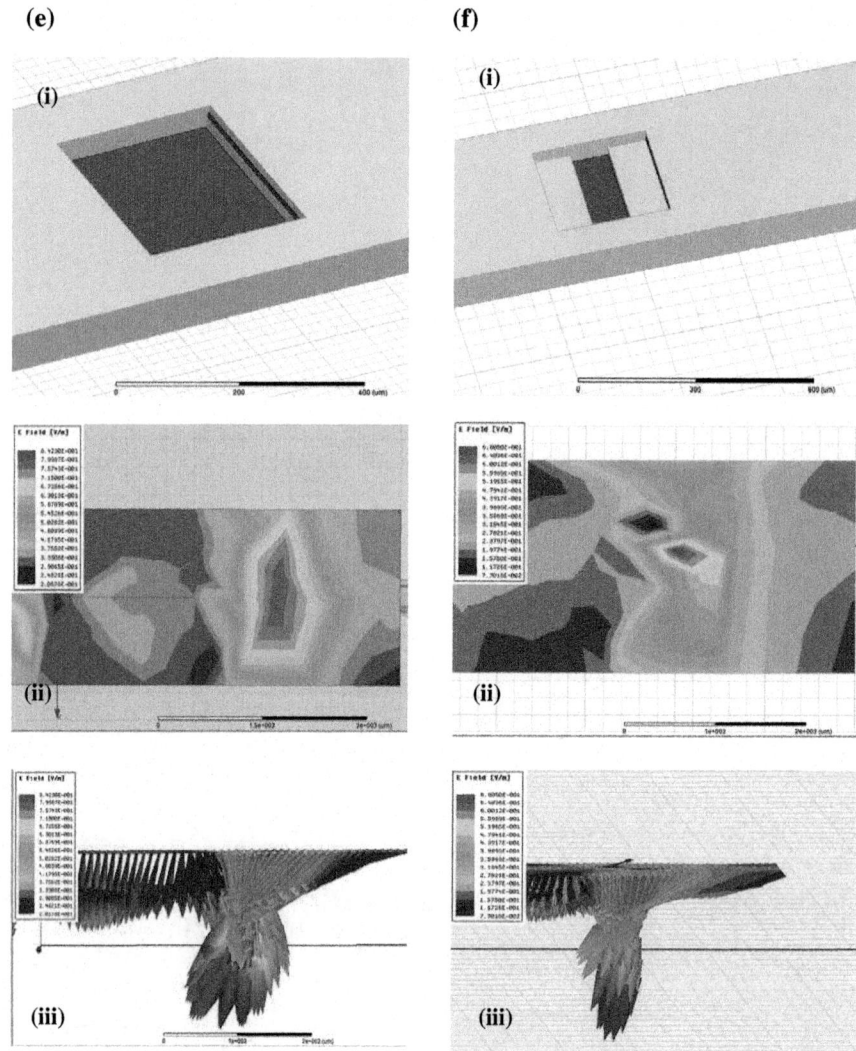

Fig. 3.14 (continued)

Table 3.1 Summarized results of absolute value of e-field distribution

	Design 0 (MP)	Design 0 (TP)	Design 1-1	Design 1-2	Design 2	Design 3-1	Design 3-2	Design 3-3
Max. E-field	4.51×10^{-7}	2.00×10^{-1}	6.87×10^{-1}	1.80×10^{-1}	8.50×10^{-1}	9.36×10^{-2}	8.46×10^{-1}	6.97×10^{-1}
1-mm-Left	2.77×10^{-7}	4.86×10^{-2}	2.26×10^{-1}	8.33×10^{-2}	3.60×10^{-1}	4.76×10^{-2}	3.20×10^{-1}	2.02×10^{-1}
1-mm-right		9.98×10^{-2}	4.13×10^{-1}	1.27×10^{-1}	3.97×10^{-1}	4.45×10^{-2}	4.78×10^{-1}	2.40×10^{-1}
500-μm-left	3.16×10^{-7}	1.02×10^{-1}	2.95×10^{-1}	1.54×10^{-1}	6.76×10^{-1}	6.70×10^{-2}	5.59×10^{-1}	2.49×10^{-1}
500- μm-right	3.31×10^{-7}	1.53×10^{-1}	5.55×10^{-1}	1.47×10^{-1}	6.65×10^{-1}	6.85×10^{-2}	6.59×10^{-1}	5.13×10^{-1}

Table 3.2 Summarized results of relative ratio of e-field distribution

	Design 0 (MP)	Design 0 (TP)	Design 1-1	Design 1-2	Design 2	Design 3-1	Design 3-2	Design 3-3
Max. E-field	1	1	1	1	1	1	1	**1**
1-mm-Left	0.615	0.243	0.329	0.463	0.424	0.508	0.378	**0.290**
1-mm-right		0.499	0.601	0.703	0.467	0.475	00566	**0.344**
500-μm-left	0.701	0.509	0.429	0.856	0.795	0.716	0.661	**0.357**
500-μm-right	0.734	0.765	0.808	0.815	0.783	0.731	0.780	**0.736**

3.3.1.3 Accelerated Soak Test and Lifetime Estimation

Leakage current measurements and MTTF of LCP-based electrode test samples using mechanical interlocking at the interface between the LCP and the gold during the lamination process and LCP-based electrode test samples without interlocking patterns are shown in Fig. 3.22. Threshold level of 1 μA indicating that device failure is determined from the earlier accelerated soak test results. The average days of soaking when the device failure occurred due to water penetration into the exposed electrode site are observed as 224 days in group 1 and 185 days in group 2, respectively. If we apply '10-degree rule', the estimated lifetimes of two groups are about 8.55 years and 7.06 years, respectively. However, there is less significant difference in data ($p > 0.05$).

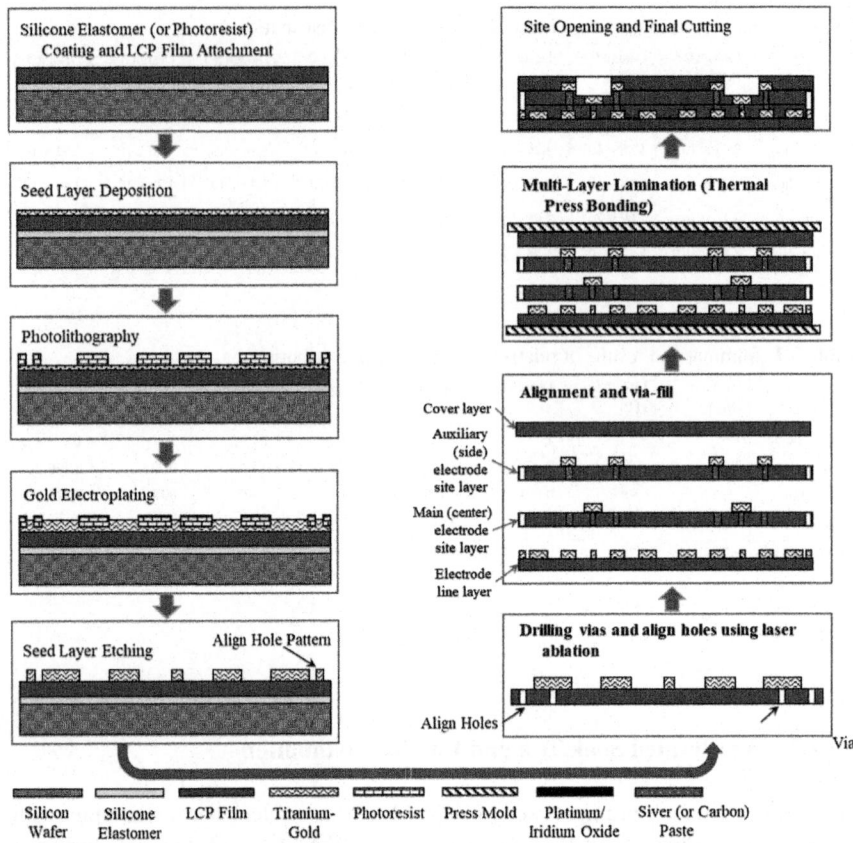

Fig. 3.15 Fabrication process of LCP cochlear electrode array for locally tripolar stimulation using MEMS technologies, laser micromachining, and thin-film processes

3.3.2 Fabrication Method Using LCP and Dielectric Materials

3.3.2.1 Adhesion Strength of Bonding Between LCP and Dielectric Materials

There are significant differences in bonding strength among LCP-LCP, LCP-SiO$_2$, and LCP-SiN$_x$ ($p < 0.05$). A shown in Table 3.5, LCP-LCP interface shows weaker bonding than LCP-dielectric materials under the low-temperature, low-pressure condition. This result reveals feasibility of dielectric material bonded to LCP.

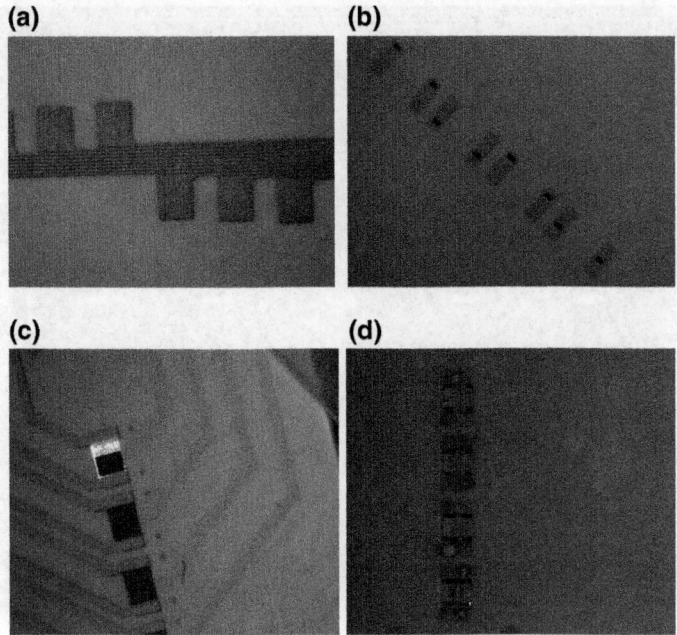

Fig. 3.16 Results of fabrication process. **a** Metal patterning after wet etching of seed layer. **b** Via opening using laser ablation. **c** Laser micromachining of site opening. **d** Opened electrode site showing center and side electrodes

Table 3.3 Comparison of reduction ratio of voltage amplitude at 1 mm distant from the center electrode

Stimulation method	Reduction ratio (%)	Voltage (V)	Reduction ratio (%)
	Left	Center (stimulation)	Right
Monopolar	–	−0.32	37.5
Tripolar (conventional)	28.57	−0.28	28.57
Locally tripolar (proposed)	39.29	−1.12	42.86

3.3.2.2   Lamination Result of the Proposed Fabrication Method

In Fig. 3.23, results of lamination process using LCP and dielectric materials on which metal is patterned are shown. There is no metal migration on the dielectric material. Dielectric materials protect metal patterning from high pressure and temperature as well as water penetration.

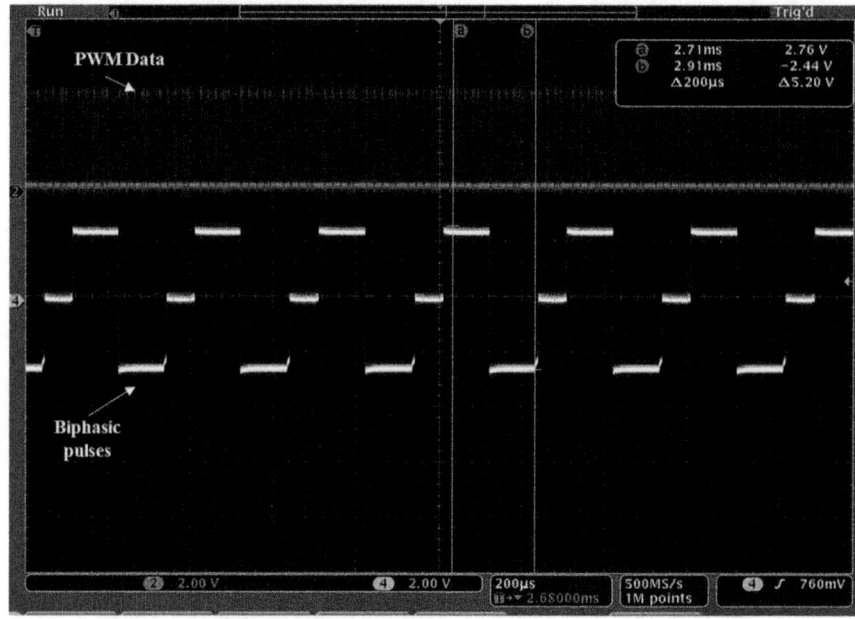

Fig. 3.17 PWM data and biphasic pulses generated by ASIC chip and test board

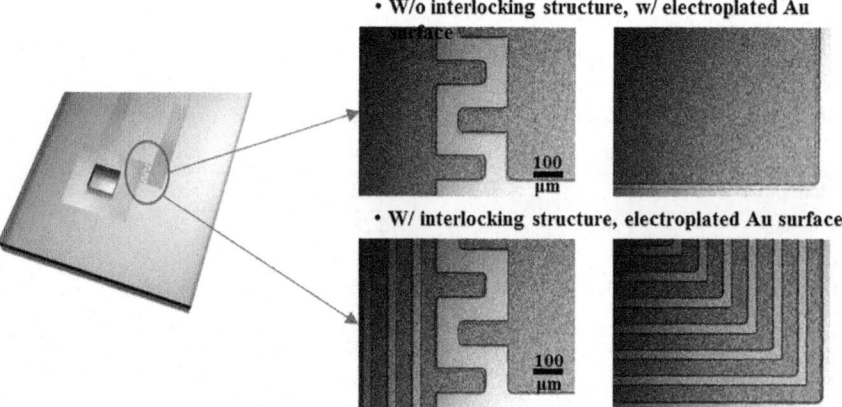

Fig. 3.18 Optical images of Electrode site and IDE pattern in fabricated electrode array with and without mechanical interlocking pattern

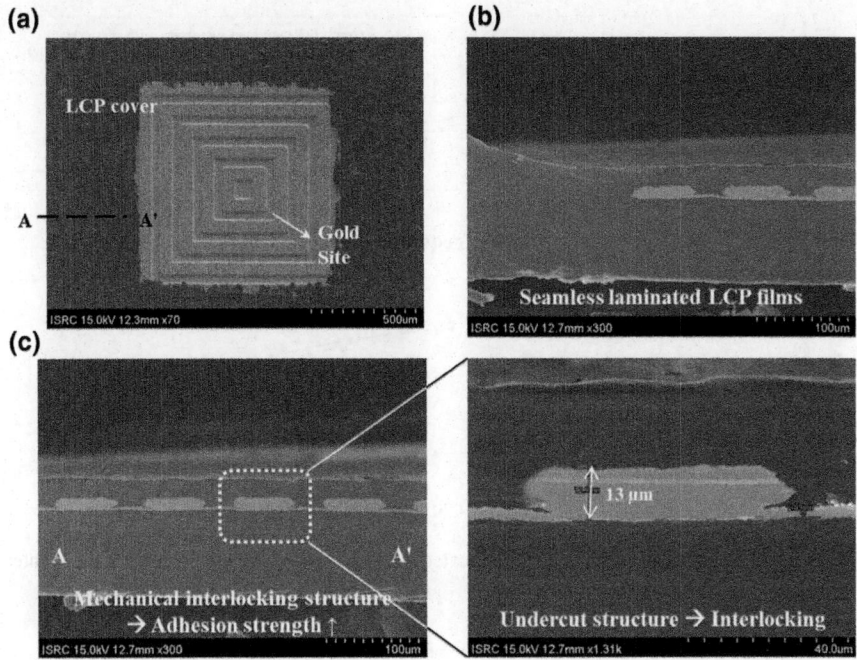

Fig. 3.19 SEM images of fabricated electrode array with mechanical interlocking. **a** Electrode site opening. **b** Seamless lamination result of upper and substrate layer of LCP. **c** Mechanically interlocked LCP film with undercut metal pattern

Table 3.4 Results of peel tests in test samples with and without mechanical interlocking

Classification	Maximum force/width (N/mm) Mean (SD)	Mode of failure
W/interlocking	0.7696 (0.0230)	Peel + snap or tear
W/o interlocking	0.5707 (0.0131)	Snap, peel + snap

Table 3.5 Results of customized peel tests to compare bonding strength of LCP-LCP with LCP-dielectric materials

Interface	Maximum force/width (N/mm) Mean (SD)	Mode of failure
LCP-LCP	0.2365 (0.0212)	Peel
LCP-SiO$_2$	0.5412 (0.0446)	Peel
LCP-SiN$_x$	0.4027 (0.0369)	Peel

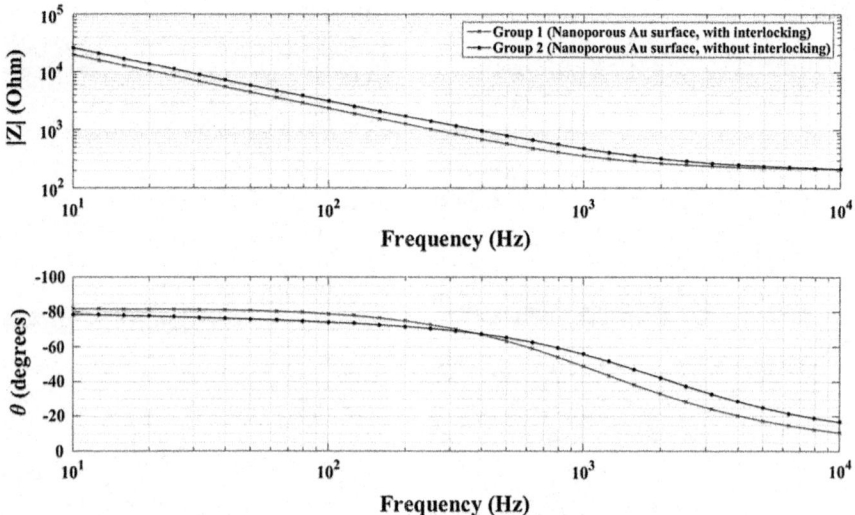

Fig. 3.20 Impedance measurements of fabricated test samples with and without mechanical interlocking used in accelerated soak test

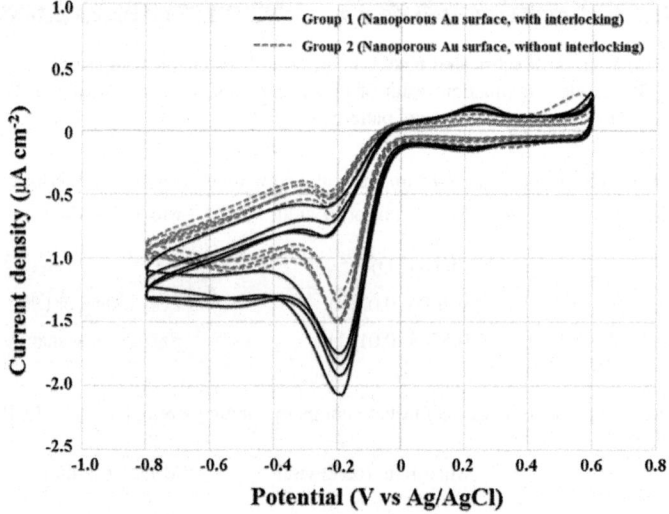

Fig. 3.21 Cyclic voltammogram to calculate charge storage capacitance (cathodic) of group 1 and group 2

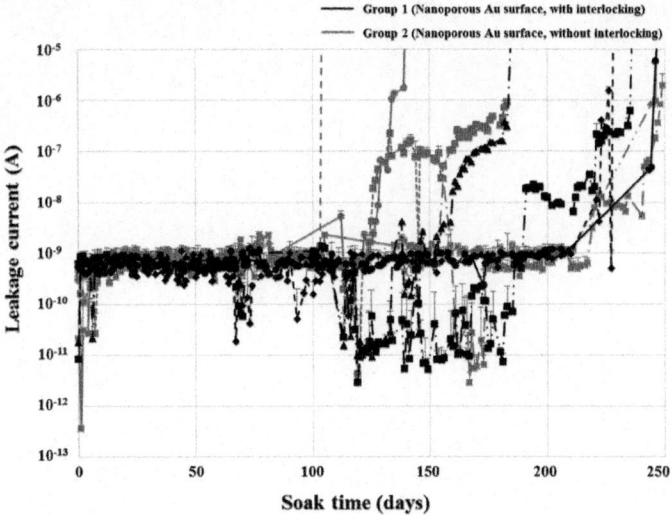

Fig. 3.22 Leakage current measurements during the in vitro 75 °C accelerated soak tests

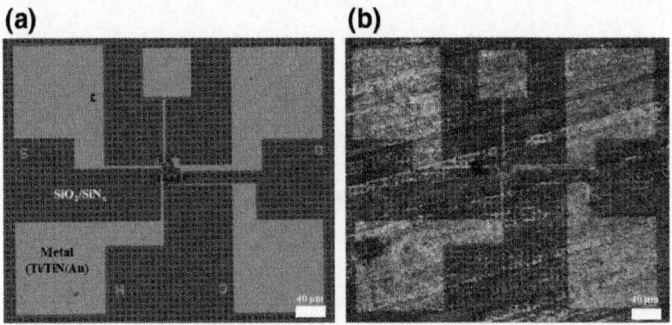

Fig. 3.23 Metal patterning **a** Before and **b** After lamination process using dielectric material and LCP

Chapter 4
Discussion

4.1 LCP-Based Cochlear Electrode Arrays for Atraumatic Deep Insertion

4.1.1 Comparison of the Current Proposed Electrode Array to the Previous Electrode Array

The proposed cochlear electrode array has a low width while maintaining variable stiffness owing to peripheral blind vias and a tapered structure. Figure 4.1 shows a comparison between the current design and the previous LCP cochlear electrode array, which had a uniform thickness throughout. If a 20-μm line pitch, 0.3-mm-wide electrode site, 16-channel electrode array, with a 0.1-mm laser machining error is assumed, the width of the previous design is 0.5–1.02 mm from the tip to the base, and the width of the proposed design is 0.3–0.36 mm from the tip to the base. The tip and base of the current electrode array can be reduced to 60 and 40% of those of the previous design, respectively (Fig. 4.2). This reduced width and more flexible tip decrease the insertion and extraction forces. As shown in Fig. 4.3, the measured insertion forces of the electrode arrays are 2.4 mN (current version), 8.2 mN (previous version), and 17.2 mN (wire-based cochlear electrode array) at a displacement of 8 mm from the round widow. The current cochlear electrode array is inserted 3 mm deeper than the previous cochlear electrode array. This means a lower frequency area can be stimulated using the current design. According to the Greenwood function, a 3-mm difference in insertion depth is approximately converted into the displacement of the array tip to the frequency region of 4009–6592 Hz, if the least-inserted array is at 20 mm [1]. As shown in Fig. 4.4, the measured maximum extraction forces are 33.0 mN (current version), 110.4 mN (previous version), and 158.5 mN (wire-based cochlear electrode array). In addition to the reduced insertion force, the extraction force of the current version is lower than that of the previous LCP-based array and the conventional wire-based array. These results are desirable for residual hearing and

© Springer Nature Singapore Pte Ltd. 2018
T. M. Gwon, *A Polymer Cochlear Electrode Array: Atraumatic Deep Insertion,
Tripolar Stimulation, and Long-Term Reliability*, Springer Theses,
https://doi.org/10.1007/978-981-13-0472-9_4

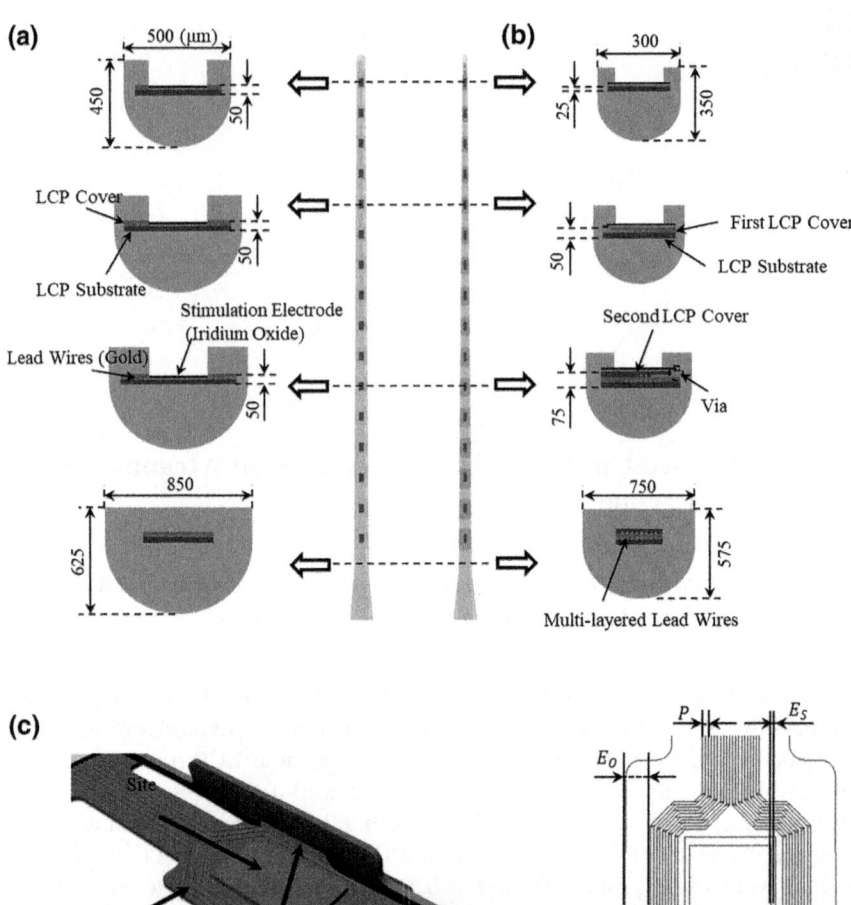

Fig. 4.1 Comparison of LCP-based cochlear electrode arrays between the previous and the current designs. **a** The thickness of the previous LCP structure is 50 μm along its entire length. Its diameters at the tip and at the base are 0.5 and 0.8 mm, respectively, and its length was 28 mm. **b** The thickness of the current LCP structure in the cochlear electrode array varies from 25 μm (tip) to 75 μm (base). **c** Previously reported array employed a design scheme which arranges stimulation sites at the center with the lead wires around the stimulation sites. **d** In this design, a peripheral via is employed so as to protect the lead wires at the center from the cutting laser beam

(d)

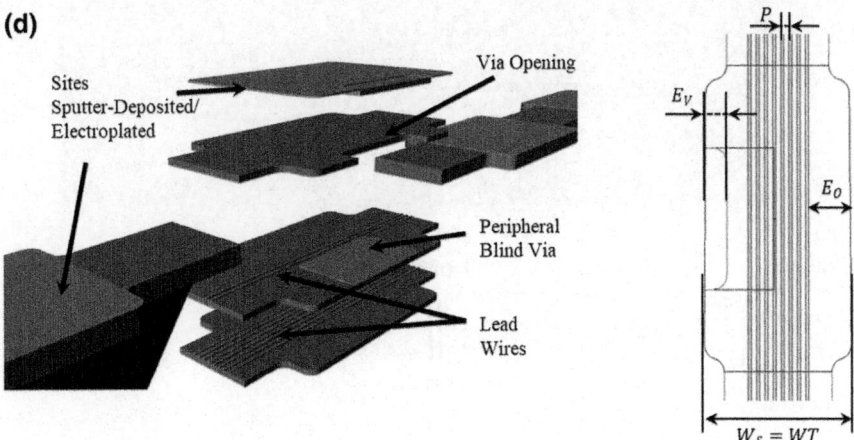

Fig. 4.1 (continued)

reimplantation surgery without insertion trauma, which is critical for children who have experienced device failure (due to the short lifespan of the device compared to the length of time these patients require the device) [2–5]. Moreover, it is possible that chemical or dry etching and laser micromachining processes can produce thinner LCP films, which would lead to a more flexible tip.

4.1.2   Improving Electrode Design Related to Insertion Depth and Trauma

The force needed for insertion into the scala tympani increases as the insertion depth increases because increased friction necessitates greater insertion force [6]. If the mechanical tolerance level of the basilar membrane is exceeded, there will be trauma to the cochlea. Uniform thickness along the entire length leads to insufficient supportability in the basal region for full insertion. In the present study, the number of LCP layers varies from the tip to the base. For a more flexible tip, channel 1 is not covered by another LCP layer, and the exposed plane is used to stimulate electrode contact. From channels 2–8, the substrate layer and intercovered layer are used for the patterns and the cover, respectively. From channel 9, intercovered layer is used for the patterning of the electrodes and wires, and the outermost insulation layer is laminated onto this layer. The more flexible tip and stiffer base result in less insertion trauma and more insertion depth, as well as deeper insertion with less insertion force.

The insertion trauma and depth of human temporal bone insertions are investigated to evaluate the proposed LCP-based cochlear electrode array with a layered and tapered design. In the present study, the measured insertion depths of the elec-

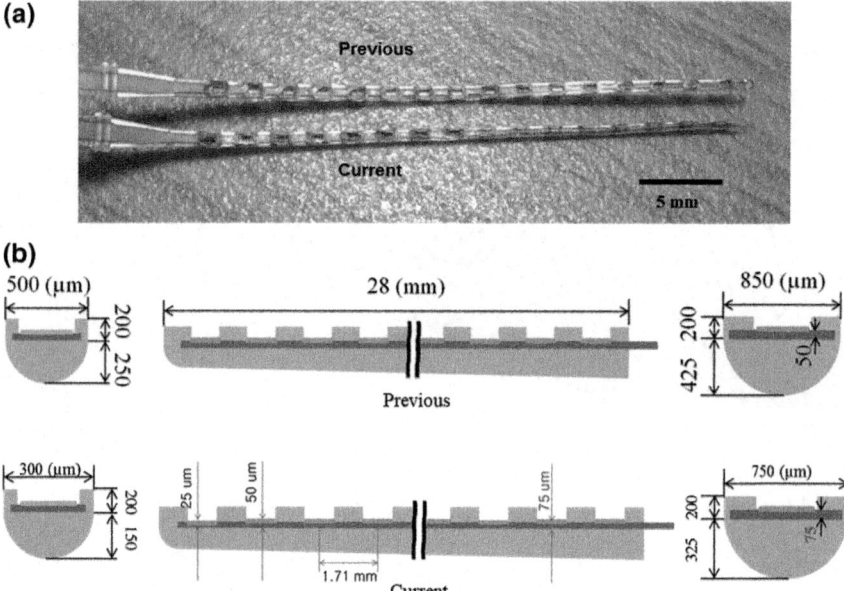

Fig. 4.2 a Optical images of fabricated LCP-based cochlear electrode arrays with and without tapered structure. **b** Side and cross-sectional views of the arrays with and without peripheral blind via

trode arrays in the human temporal bone are 630°, 470°, 450°, and 360° in the round window approach and 500° in the cochleostomy approach (Fig. 4.5). Four cases show enhanced results of the insertion depth, but for one case, the insertion length was shorter than the previous result (405°). Insertion trials generally encounter considerable resistance from the 13th electrode site insertion of 16. There is no significant difference between the round window approach and the cochleostomy approach in terms of resistance. The insertion trauma continues to exist only in the 630° temporal bone insertion trials, with evidence that the electrode array causes no trauma in the basal turn. In the 630° insertion trial, the LCP-based cochlear electrode array is found to be wavy from the result of treatment and handling. Therefore, the wavy structure in the electrode substrate may help reduce the insertion force because film-type LCP-based cochlear electrode arrays undergo resistance perpendicular to the plane of the LCP surface. Although atraumatic deep insertion is observed with the proposed finer and multi-layered electrode array, additional approaches, such as a more flexible tip and a wavy substrate, are needed to insert the LCP-based cochlear electrode array deeper without trauma in the basal and second turn.

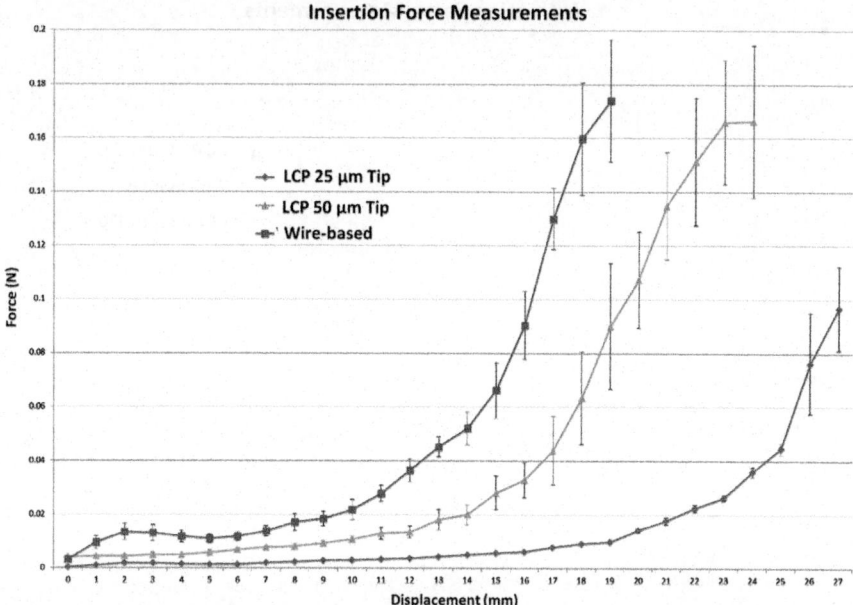

Fig. 4.3 Insertion force measured in a customized insertion setup (25 and 50 μm tip mean the current design and the previous design, respectively)

4.1.3   Aspects to Improve in the Fabrication Process

Multi-layered structures can be also used to provide a manufacturing method for high-density, implantable, polymer-based electrode arrays. Using a multi-layered structure and a microfabrication process allows more electrode stimulation channels (given the restricted electrode dimension). This is not easily achieved in a manual, wire-based cochlear electrode array. The more the LCP layers, the more usable stimulation channels. This can be another advantage of a cochlear electrode array, which has a limited number of stimulation channels compared to the number of hair cells, until the LCP layers are too stiff to be inserted stably.

Figure 4.6 shows the fabrication processes and their results. In the lamination process to transform the multi-layered films to one united substrate, alignment and even pressure are significant factor for determining the success in the fabrication of the LCP electrode array. As in Fig. 4.7, metal lines in the bottom layer moved to upper layer because of the stepped pulley structure shown. This often causes short or open metal lines, resulting in reduced production throughput. Partially stepped structures are inevitable if the via structure and thermally compressive lamination process are used.

For atraumatic insertion, it is also important to produce a fine electrode array. Fine electrode arrays cause less damage to the cochlear tissue. To reduce the width of the electrode array, fine metal patterns are useful. Because the metal is evaporated and

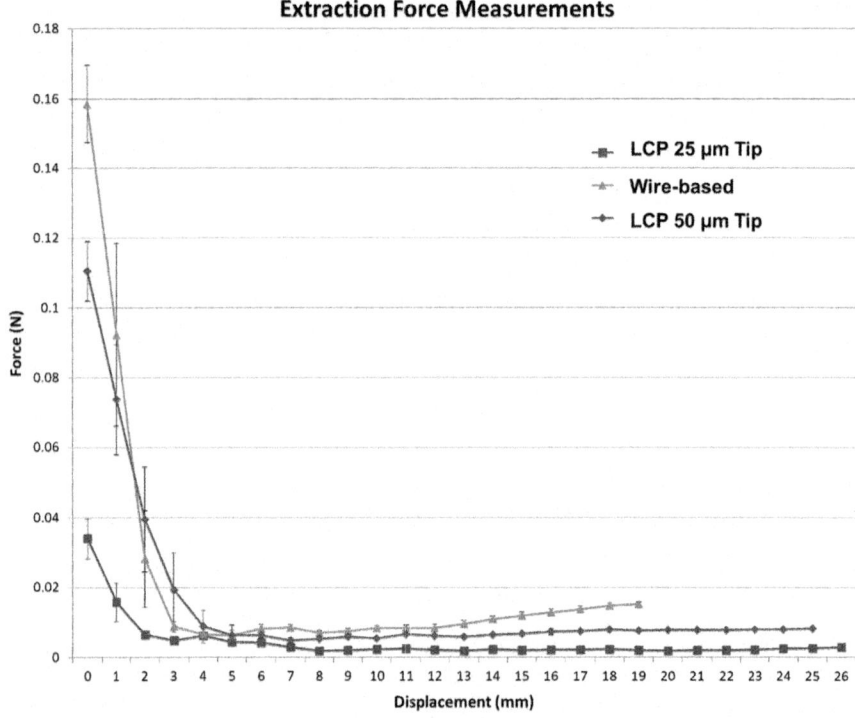

Fig. 4.4   Extraction force measurements

patterned onto LCP films on silicon wafers, the attached LCP must lie uniformly flat on the silicon wafer. Therefore, it is necessary to use proper adhesive materials to attach the LCP film onto the silicon wafer. Conventionally, a photoresist has been used to attach LCP films to silicon wafers. In microfabrication processes, the evaporation of the solvent on the photoresist occurs. The flatness of the LCP film attached using the photoresist layer worsens over time, and this method cannot support fine metal patterns throughout the manufacturing process. Specifically, equipment using a vacuum chamber creates more severe situations, resulting in a limitation in terms of the dimensions of the metal pattern. In contrast to a photoresist, a silicone elastomer does not create bubbles or voids. Moreover, it is resistant to chemical solutions. In addition, the detachment of the LCP films should be considered for micropatterning. The LCP film should adhere to the wafer during the process with a chemical etchant and solvent. If an adhesion layer is damaged or dissolved in these chemicals, the microstructure cannot be patterned onto the LCP film. The degree of bumpiness of the LCP films attached to a silicon wafer using a photoresist and a silicone elastomer is greater than 43.7 and 16 μm, respectively [7]. Because the size of the electrode array is related to the degree of trauma, it is important to use a uniform, flat adhesion layer when manufacturing LCP-based electrode arrays. The silicone elastomer shows better uniformity and durability than a photoresist. Unless peeling

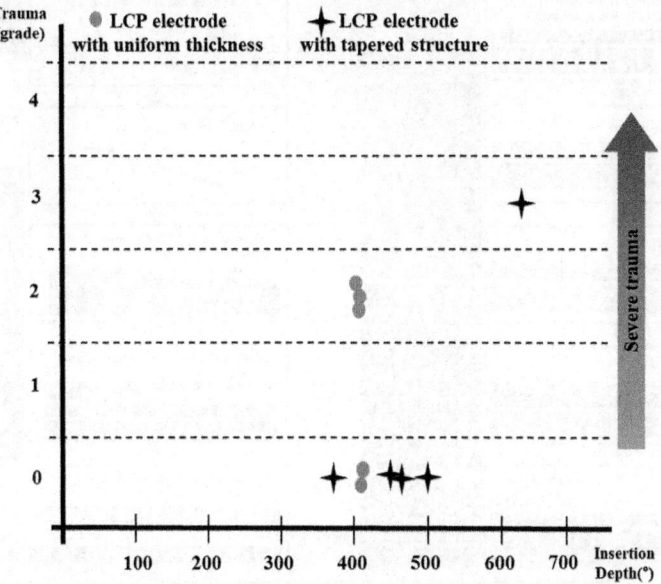

Fig. 4.5 Insertion depth and trauma in human temporal bone insertion studies of LCP-based cochlear electrode arrays

stress is applied, the film adhered to the silicone elastomer is not detached. Because the Van der Waals force [8, 9] is applied between the silicone elastomer and LCP film, the elastomer-coated silicon wafer can be reused for additional microfabrication processes.

4.1.4 In Vivo Implantation

As shown in EABR recordings, electrical stimulation was successfully delivered to the cochlea of a guinea pig through the current electrode array. Acute insertion and stimulation steps confirm its functional operation as an electrode. In future work, it is necessary to implant the entire LCP-based cochlear implant system into an animal. Chronic electrical stimulation is required to analyze the long-term reliability and histological changes of the device.

Electrode implantation was maintained stably during the four-week implantation of dummy LCP electrode arrays in guinea pigs. Tissue reaction was comparable to that of conventional cochlear electrode array implantation. The scala tympani area occupied by tissue responses were 52% control group (Cochlear™ electrode array) and 43% experimental group (LCP electrode array). Figure 4.8 shows fibrosis in both electrode array implantations. The electrode array was implanted in the scala

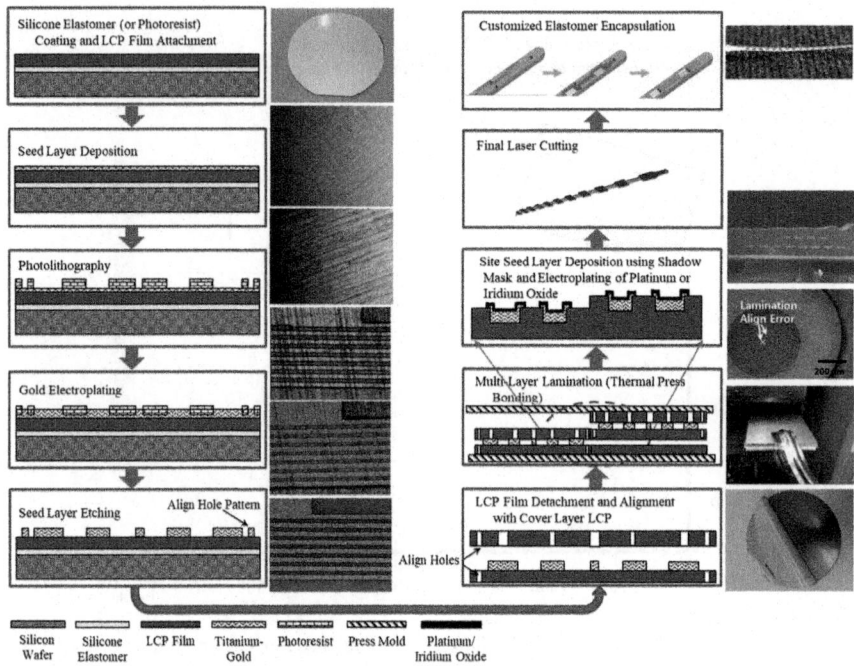

Fig. 4.6 Fabrication processes and their results

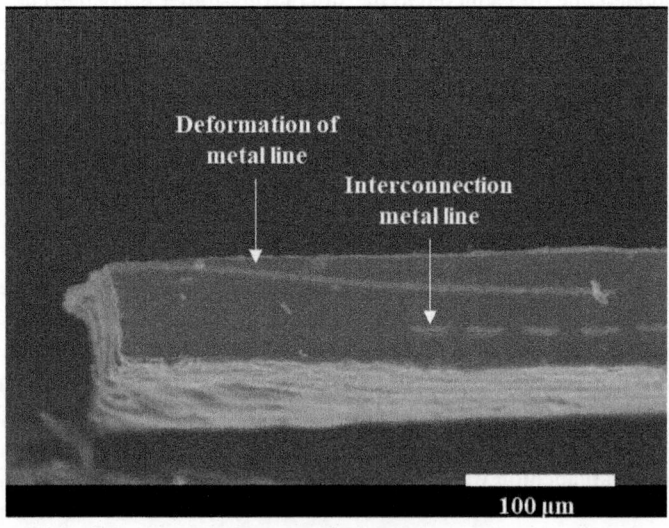

Fig. 4.7 Deformation of metal line after lamination process

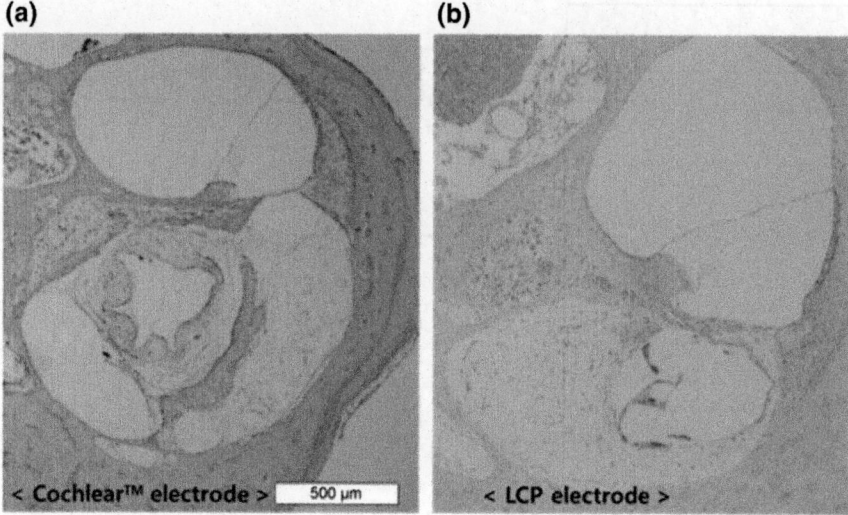

Fig. 4.8 Comparison of fibrosis between conventional and LCP-based cochlear electrode array

tympani and fixed at its position and there was no significant difference between the two images. Feasibility for clinical use of LCP-based cochlear electrode arrays is corroborated by these in vivo implantation results.

4.2 Power Consumption and Stimulation Threshold of Tripolar Stimulation

Tripolar stimulation for current focusing effectively reduces crosstalk between electrode stimulation channels while increasing power consumption [10–12]. Multipolar strategies utilize concurrent stimulation of multiple electrode sites, resulting in increased power consumption. However, high threshold and comfort levels can decrease the time for which batteries can be used [10, 13]. Loudness (amplitude of the stimulation current pulse) can be limited using compliance voltage owing to an increased threshold level. Larger channel-to-channel variability in the stimulation level is another consideration from the focused current and reduced reactive areas. There are alternatives to the trade-off of reduction in channel interaction and power consumption, such as reducing the stimulation pulse rate, reducing the distance from the electrode to target tissue, and limiting the number of channels in use. Paired stimulation (depicted in Fig. 4.9), which is stimulation using two electrode sites simultaneously, can cut the overall stimulation rate by half [12]. This method is interrupted in practical application by channel interaction. Convergence tripolar stimulation and paired stimulation can balance each other in terms of the increased power consumption and channel interaction. If the proposed electrode site designed

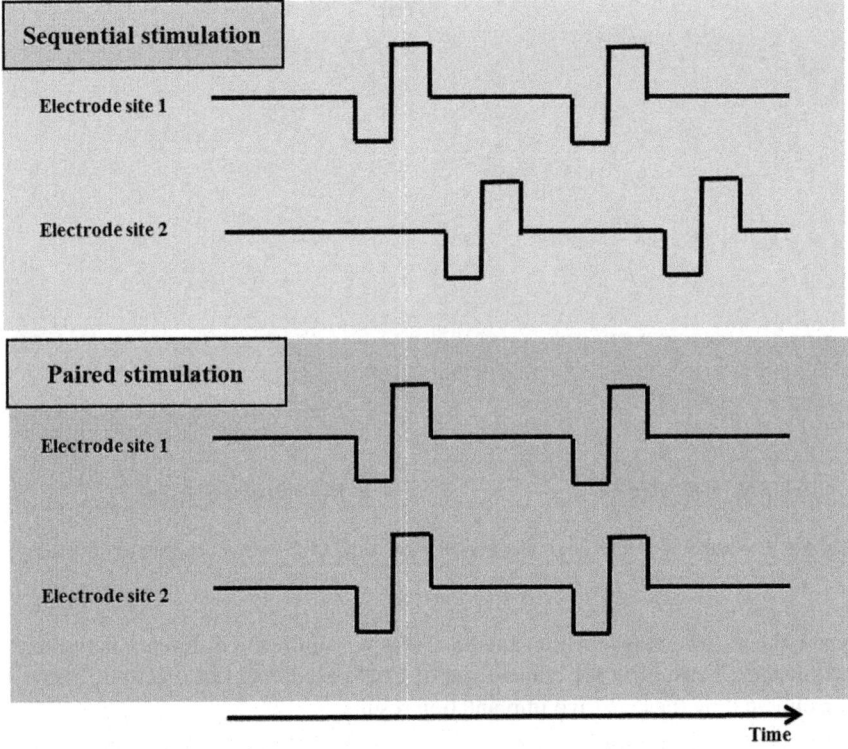

Fig. 4.9 Scheme of electrical pulse trains of sequential and paired stimulation

for local tripolar stimulation adopts paired stimulation, adjacent electrode sites can be used in stimulation simultaneously, which physically increases the number of stimulation channels in the tripolar stimulation while power consumption is halved due to the reduction of stimulation rate.

4.3   Technical Strategies to Improve Device Reliability

4.3.1   *Mechanical Interlocking at the Metal-Polymer Interface*

Metallic lines in the electrode site are patterned using photolithography and electroplating to fabricate mechanical interlocking patterns. Overexposure to UV light forms trapezoidal photoresist patterns, which results in anti-trapezoidal metallic patterns during the electroplating process. Thermal lamination bonding interlocks the LCP insulation layer with anti-trapezoidal (undercut) metal structures (without forming voids) for seamless bonding between LCP films for electrode substrates and insulation LCP films.

Optimal adhesion joint strength between two materials is defined as follows [14]:

$$\text{Optimum joint strength} = (\text{constant}) \times (\text{mechanical interlocking component})$$
$$\times (\text{interfacial chemical component})$$

Interfacial chemical bonding is determined by the reactivity of the involved materials [15]. Polymers are attached to noble metals by intermolecular electrostatic and weak van der Waals forces because noble metals have no reactivity to organic polymers like LCPs [16]. In this strategy, the mechanical interlocking using microscale metallic lines for interfacial mechanical bonding is adopted to improve interfacial adhesion strength between LCPs and noble metals for long-term device reliability. In peel tests, the adhesion strengths are significant different between test samples with and without mechanical interlocking. Metal surface topography modification causes the failure mode to transition from adhesive to cohesive [17]. Both groups with and without interlocking structures have nanoporous surfaces formed by electroplating in electrode sites, which leads to failure modes in both groups that are not only adhesive. Mechanical interlocking provides the energy expending process for crack propagation, which makes the interface more cohesive. In addition to enhancing the adhesion strength, the structures remaining on the exposed area of the electrode site also increase the surface area. Compared to the control group, the impedance is 24.8% lower and CSCc is 34.6% higher owing to a larger surface area.

The size of interlocking patterns is critical to adhesion strength [17]. In the previous study, the failure load of adhesive joints increased according to the ratio of the patterned width to the preserved width. In this strategy, the size and morphology of the metallic pattern for interlocking on the polymer film need to be separately studied. However, in this study, the effect of the mechanical interlocking pattern on adhesion strength and device reliability was proved by comparing the test samples with and without mechanical interlocking, though the interlocking pattern of noble metal was not optimized.

Accelerated soak testing was used to investigate device reliability. In addition to the results of the peel tests, the results of accelerated soak tests revealed the practical reliability of LCP-based devices. In Fig. 4.10, water penetration at the metal-LCP interface caused detrimental effects in LCP adhesion, resulting in metal line breakage and device failure. However, mechanical interlocking between a noble metal and LCP was intact even after water penetration, as shown in Fig. 4.11. Both the top and bottom surfaces of the metal were peeled off from the LCP insulation film and electrode substrate in non-mechanical interlocking samples, but adhesion between the interlocked metal and LCP was maintained after water ingress. In mechanical interlocking samples, there was further delamination at the interface without interlocked structure. The interlocked interface between LCP and undercut metallic structure remained intact after device failure, while the other interface (without interlocking) was delaminated even in the same test specimen. Therefore, this result showed

Fig. 4.10 Optical microscope images of test samples. **a** Electrode site of test sample without mechanical interlocking in the course of accelerated soaking. **b** Electrode site of test sample with mechanical interlocking after water penetration. **c, d** Backlight microscope Images of test samples without and with mechanical interlocking after device failure. **e** Side view of delamination of LCP film. **f** Breakage of metal lines after water penetration

that double-sided mechanical interlocking is needed to enhance device reliability. Mechanical interlocking at the fabrication steps of metal deposition on LCP film can be realized using following fabrication process (Fig. 4.12). In Fig. 4.13, mechanical interlocked metal-LCP in the substrate (bottom) layer on which metal patterning can be fabricated.

Titanium can be used to strengthen interfacial chemical bonding as an adhesion promotion layer, however, more complicated and delicate fabrication processes are needed due to the risk of galvanic corrosion from multiple metal layers exposed to ionic water [18]. When using different metals, microcracks in the electrode site can affect device reliability. Microcracks exposed to electrolytes accelerate water penetration. Figure 4.14 shows TEM and FIB images of microcracks in the LCP-based neural electrode due to mishandling. Water ingress into microcracks can cause damage to electrode sites and galvanic corrosion at the interface of two different metals.

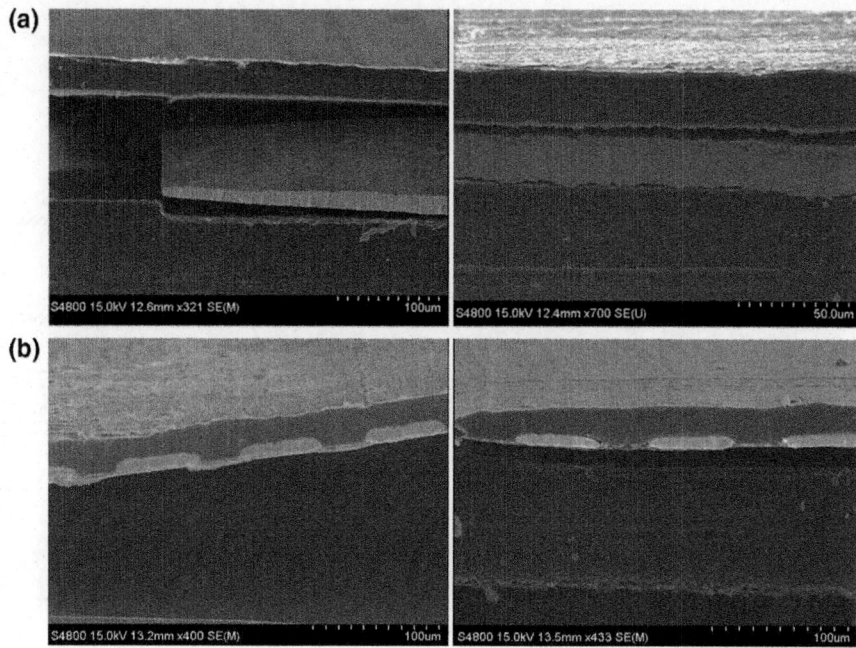

Fig. 4.11 SEM images of cross-section views of test samples after accelerated soak tests. **a** Without mechanical interlocking. **b** With mechanical interlocking

4.3.2 Hybrid Device Based on Polymer and Dielectric Materials

The results of customized peel test revealed that LCP-dielectric bonding is more reliable than LCP-LCP bonding when using low-temperature and low-pressure lamination methods, even though bonding conditions, such as plasma treatment, are not optimized. These results show feasibility of the fabrication process using dielectric materials with LCP. However, as in the results of the peel tests, the LCP-dielectric interface is adhesive from the failure mode. Conversion from adhesive interface to cohesive interface is needed for better reliability using anodic bonding and optimization of plasma conditions for surface activation.

Figure 4.15 shows the peeled interface of LCP-LCP bonding according to lamination temperature and pressure. In samples with high-temperature, high-pressure lamination processes, the film does not peel neatly, and it is difficult to distinguish the interface. In contrast, the LCP film bonded in low-temperature, low-pressure lamination is peed from bottom LCP film, and the interface is clearly distinguished. In this regard, it is essential to use high-temperature, high-pressure lamination processes for long-term device reliability. To prevent migration of metal lines, which is bad for

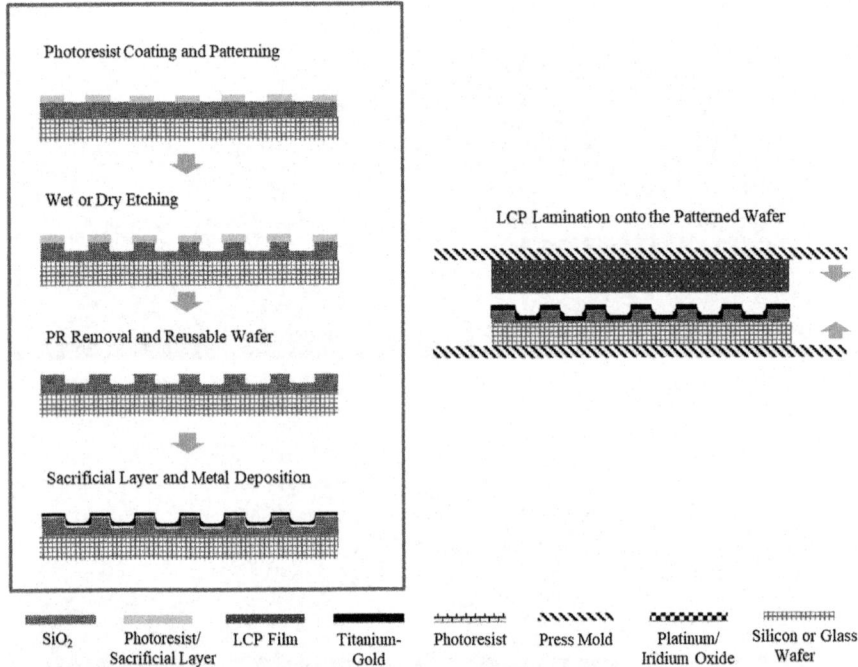

Fig. 4.12 Fabrication process of mechanically interlocked metal deposition onto LCP film (1st sacrificial layer and SiO₂ deposition on Si or Glass wafer → Interlocking patterning using photolithography and SiO₂ etching → 2nd Sacrificial layer and noble metal deposition → Lamination metal patterning onto LCP film → SiO₂ and 2nd sacrificial layer removal)

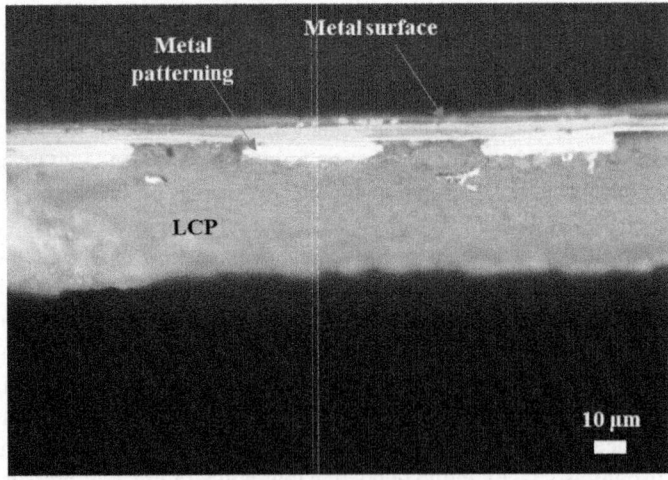

Fig. 4.13 Mechanical interlocked metal-LCP in the substrate layer

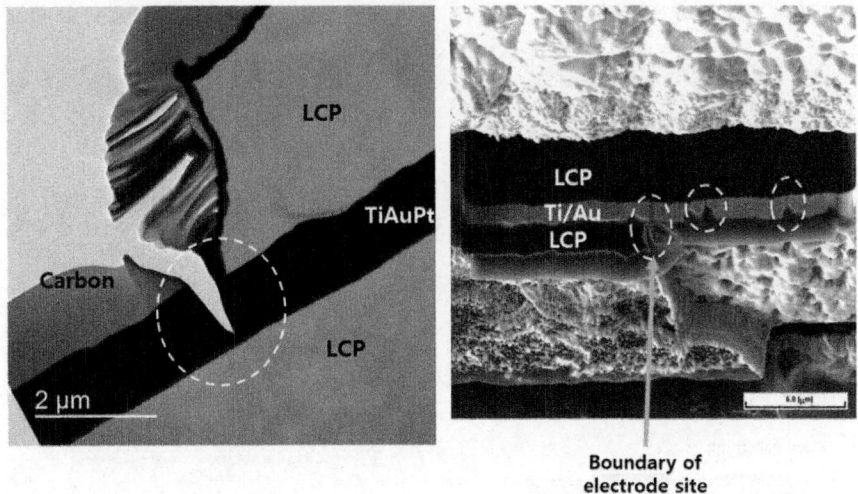

Fig. 4.14 **a** TEM and **b** FIB-SEM images of micro-crack in electrode site

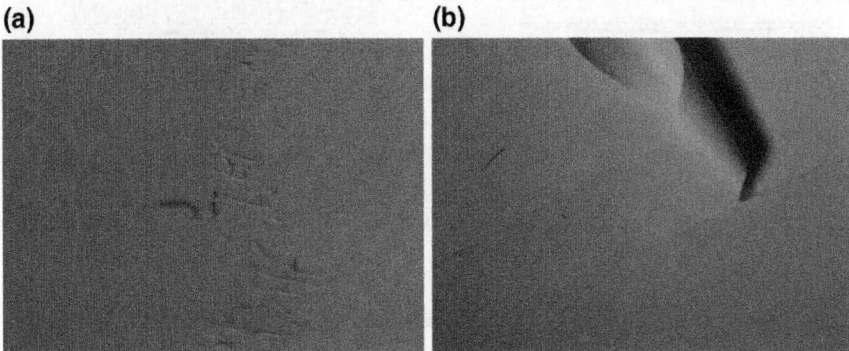

Fig. 4.15 Images of LCP-LCP interface. **a** Using high-temperature, high-pressure lamination process and **b** Low-temperature, low-pressure lamination process after peeling off

high-temperature, high-pressure lamination, dielectric materials can be utilized to grab the metal patterning during lamination process. There is no significant change in metal patterns before and after the lamination process.

Figure 4.16 shows a feasible fabrication process and metal pattern with dielectric materials on LCP film. Long-term, reliable, high-density LCP-based neural implants can be achieved using this fabrication process. It is difficult to handle layers of dielectric materials due to their thinness (<5 μm). LCP plays the role of handling substrate as well as final cover layer.

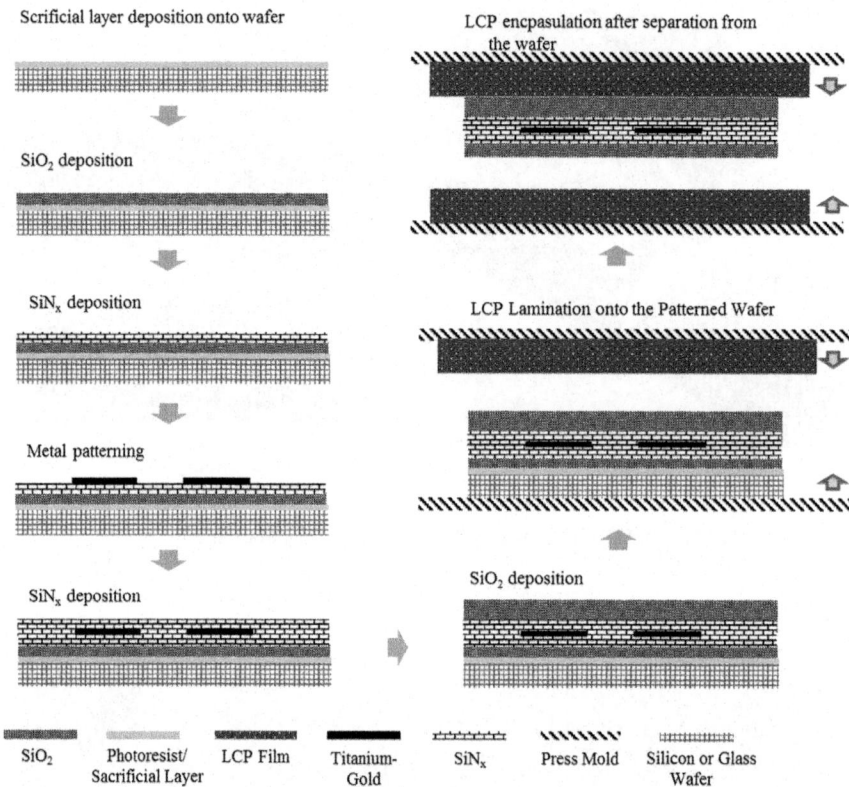

Fig. 4.16 Fabrication flow feasible to make LCP-dielectric material hybrid implantable device

4.4   Review of Long-Term Reliability of LCP-Based Devices

Helium leak tests are the standard method used to measure the hermeticity of metal-based implantable devices. The detected helium leakage in this test is estimated against that of water using the square root of the ratio of their molecular masses [19]. However, in polymer-based devices, it is difficult to measure helium leakage, as organic polymers, such as LCPs, naturally absorb helium [20, 21]. Also, in terms of the electrode, water ingress at the interface between the polymer and the metal at the electrode site cannot be separated from that at the surface of the electrode. The absolute value of long-term reliability of polymer-based implantable devices is hard to measure using the conventional testing methods for metal-based devices. Therefore, testing methods, such as accelerated soak tests, are needed to compare the "relative long-term reliability" of the test samples. As mentioned in the Chap. 2, Materials and Methods, there are three pathways through which water ingress occurs in the LCP-based neural electrodes: through the LCP surface, through the interface between the LCP and the LCP layers, and through the interface between the LCP and the metal. Among these pathways, the polymer-metal interface is the limiting factor for the long-term reliability of LCP-based neural implantable devices [22]. An examination taken after water leakage into the electrodes corroborates that the LCP-metal interface is the region most vulnerable to water penetration.

The accuracy of lifetime estimations at body temperature using the 'ten-degree rule' is not completely acceptable. The ten-degree rule is based on gross approximations from the Arrhenius reaction rate function. To obtain an accurate lifetime estimation, the activation-energy-related mechanism should be identical at the body and accelerated temperature. In this regard, an accelerated temperature above 60 °C is not recommended in a polymer-related aging test owing to the non-Arrhenius behavior of polymers during degradation [23]. Other activation energy-related mechanisms that do not occur at typical body temperatures can occur at an elevated temperature. Preliminary accelerated soak tests using LCP-based neural electrodes at 110 and 95 °C also revealed that theoretical predictions of device lifetime based on the ten-degree rule are not in agreement with empirical data. The ten-degree rule provided conservative predictions of lifetime expectancy [24]. For the soak tests, 75 °C was chosen as an elevated temperature to expedite failure within a limited period. In addition, temperatures above 60 °C have been used in accelerated aging tests for polymer-based device, such as parylene-C and LCP, which were used to prove the reliability of the fabrication methods [22, 25]. These results can be used to compare the "relative" reliability of LCP-based neural electrodes depending on the mechanical interlocking at the LCP-metal interface. To obtain more accurate lifetime estimations, an accurate temperature coefficient (Q_{10}), which is generally around 2 generally, resulting on increasing the temperature by 10 °C should be calculated through accelerated soak tests at multiple temperatures within a narrow temperature interval (such as 5–10 °C).

Electrical modeling analyzed in this dissertation refers to the literature [26] about modeling of recording electrodes and it is transformed according to stimulation electrode arrays. It is assumed that LCP-LCP bonding is strong and delamination between LCP films does not occur when using high-temperature, high-pressure lamination. However, if insufficient temperature and pressure are given due to stability to secure metal patterning, LCP-LCP and metal-LCP bonding are delaminated when water penetrates the LCP-based device. LCP-LCP delamination affects metal line stability and electrolytes penetrate into the space between metal lines, resulting in a short. This causes crosstalk between adjacent electrode stimulation sites. In the model (Fig. 4.17), crosstalk (dB) is calculated as follows:

$$\text{Crosstalk (dB)} = -20 \log \left(\frac{V_{in1}}{V_{out2}} \right)$$

$$\approx -20 \log \left(\frac{1 - \frac{(Z_{site} \| Z_{c,site})(Z_{sh} \| Z_{c,line} \| Z_{site})}{Z_{site}^2} - \frac{(Z_{site} \| Z_{c,site})^2}{Z_{c,site}^2}}{\frac{(Z_{site} \| Z_{c,site})(Z_{sh} \| Z_{c,line} \| Z_{site})}{Z_{site} Z_{c,line}} + \frac{(Z_{site} \| Z_{c,site})^2}{Z_{site} Z_{c,site}}} \right)$$

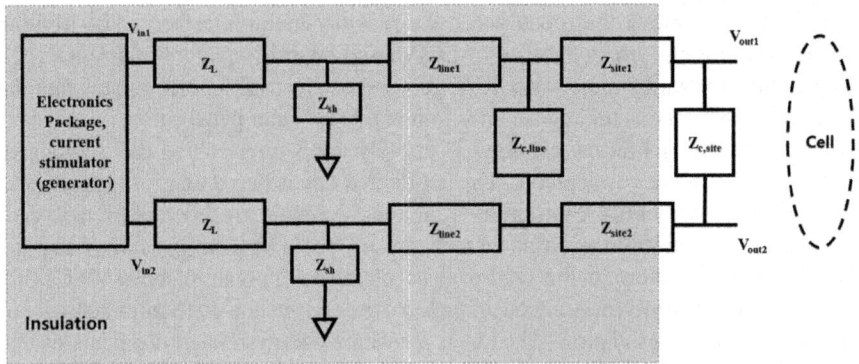

Z$_{site}$: electrode impedance
Z$_c$: coupling impedance, crosstalk impedance
Z$_{line}$: line impedance
Z$_L$: output impedance, load impedance
Z$_{sh}$: wire shunt or parasitic impedance to ground

Fig. 4.17 Electrical modeling of crosstalk analyzation in stimulation electrode when water penetration occurs

It is assumed that Z_L and Z_{line} are negligible and Z_{sh} is much larger than other impedances. The pitch of metal lines is 20 μm and the conductivity of PBS is 15,000 μΩ^{-1}/cm. Volume resistivity of LCP is larger than 10^{13} Ω cm. The area of the electrode site and distance between electrode sites are 0.1 mm^2 and 1 mm, respectively. In high-temperature, high-pressure lamination, $Z_{c,line}$ is approximately equal to R_{LCP}, which is 1.33×10^{15} Ω. In low-temperature, low-pressure lamination, $Z_{c,line}$ is approximate to R_{PBS} after water ingress, which is 8.89×10^3 Ω. This low value of $Z_{c,line}$ increases the crosstalk value, which leads to device failure. Therefore, high-temperature, high-pressure lamination process is important for device reliability.

The results of reliability tests for various LCP-based devices are summarized in Table 4.1. To predict the relative humidity inside the LCP-encased cavity, the moisture diffusion through the polymer was calculated using Fick's diffusion law and Henry's law with solubility [22]. The theoretical lifetimes (which were proportional to the thickness of the LCP layers because of the time constants when the relative humidity reaches 63 of % inside the LCP-based package) for LCP layers 25, 50, and 100 μm thick, were 15.3, 31.2, and 64.3 years, respectively.

In a study to compare an LCP-based implantable package with polyimide- and parylene-based packages using IDEs, the leakage current was not detected during more than one year of soaking in 75 °C PBS [27]. In contrast, other polymer-based devices failed within 117 days. The leakage current in a retinal prosthesis with a curved package design was also measured [22]. After 87 days of soaking at 87 °C PBS, water penetration occurred.

Ion diffusion and oxidation are the main causes of the delamination of polymer–metal interfaces in ion-containing saline [32–34]. Lee et al. evaluated the adhesion of LCP–LCP and LCP–Ti interfaces in 37 °C PBS soak tests which incorporated

Table 4.1 Summarized results of reliability tests of LCP-based devices and the conditions of each specimen

Classification	Interface	T^a (°C)	MTTF^b (days) or leak rate	Lamination pressure (kPa)	Surface modification and bonding	Method
LCP encapsulation [22]	LCP surface	Body	>5000	–	25-μm thick LCP	Analytic calculation
Flat package [27]	LCP-LCP	75	390	<100	O$_2$ plasma fusion bonding	Soak (IDE)
Curved package [22]	LCP-LCP	87	87	500	O$_2$ plasma fusion bonding	Soak (IDE)
Electrode [22]	LCP-metal	87	21	75	O$_2$ plasma, electroplated gold fusion bonding	Soak (pulsing)
Electrode [22]	LCP-metal	87	114	500	O$_2$ plasma, electroplated gold fusion bonding	Soak (pulsing)
Electrode [28]	LCP-metal	75	185	100	O$_2$ plasma, electroplated gold fusion bonding	Soak (IDE)
Electrode [28]	LCP-metal	75	224	100	O$_2$ plasma, electroplated gold, interlocking structure fusion bonding	Soak (IDE)
Flexible package [29]	LCP-LCP	Body	42	–	Fusion bonding	In vivo
Flexible package [29]	LCP-LCP	75	95	–	Fusion bonding	Soak (gate-drain leakage current)
Interconnection [30]	LCP-metal	–	<5 × 10^{-8} (mbar · l/s)	–	Fusion bonding	Helium leak
Surface-mount package [20]	LCP-LCP	–	3.7 × 10^{-8} (atm · cc/s)	–	Fusion bonding	Helium leak

^{a}Temperature used in the soak tests
^{b}Mean time to failure
Reprinted from the Ref. [31]

blister tests [35]. The test results showed that LCP–LCP and LCP–Ti interfaces were more reliable than polyimide–polyimide and polyimide-Ti interfaces. Methods to enhance the adhesion strength between LCPs and metals have been explored, and their effects on reliability were evaluated [22, 28]. A device created at a thermo-compression pressure of 500 kPa withstood operation 93 days longer (from 21 to 114 days) than another created at 75 kPa of pressure in 87 °C PBS. Another method for enhancing the adhesion strength involved the use of mechanical interlocking between an LCP and metal [28]. Micro-metal patterns for mechanical interlocking of LCP with metal were attached to LCP substrates. The LCP insulation layer and the metal patterns were interlocked during the thermo-compression bonding process. The test specimen with interlocking showed higher adhesion strength and improved device reliability, increasing from 185 to 224 days, in a 75 °C soak test.

In vivo device stability tests for LCP-based packaging technologies have been performed. An LCP-encapsulated wireless system was implanted in a rat for six weeks to verify the RF response stability of the device that the external PCB linked to a network analyzer. Insertion loss and isolation remained constant for six weeks [29]. Non-functional dummy LCP devices for retinal prosthesis were implanted stably in two New Zealand white rabbits for 2.5 years without adverse effects or structural deformations [22, 35].

Although it has been argued whether helium leak tests can be used for polymer-based packages, there have been several helium fine-leak tests of LCP coatings. High-density electrical interconnection feedthroughs (1024-stimulator channels in a 5 mm × 5 mm area) were realized using a novel fusion bonding process [30]. Helium leak rates of $<5 \times 10^{-8}$ mbar $\cdot$ l s^{-1} were measured, which is a reliable level compared to the leak rates of glass substrates with metallized vias. The LCP cavity of LCP surface-mount packages for Kα-band applications had a fine leak rate of 3.6×10^{-8} atm $\cdot$ cc s^{-1} [36]. Another LCP-based package using the same thermo-compression sealing process showed a measured fine leak-rate of 3.7×10^{-8} atm $\cdot$ cc s^{-1}. These results show that an LCP is classified as a near-hermetic material for electronics packaging comparable to ceramics and metals [20].

References

1. D.D. Greenwood, A cochlear frequency-position function for several species—29 years later. J. Acoust. Soc. Am. **87**, 2592–2605 (1990)
2. S.J. Rebscher, A. Hetherington, B. Bonham, P. Wardrop, D. Whinney, P.A. Leake, Considerations for design of future cochlear implant electrode arrays: electrode array stiffness, size, and depth of insertion. J. Rehabil. Res. Dev. **45**, 731–747 (2008)
3. R. Shepherd, K. Verhoeven, J. Xu, F. Risi, J. Fallon, A. Wise, An improved cochlear implant electrode array for use in experimental studies. Hear. Res. **277**, 20–27 (2011)
4. S.C. Parisier, P.M. Chute, A.L. Popp, G.D. Suh, Outcome analysis of cochlear implant reimplantation in children. The Laryngoscope **111**, 26–32 (2001)
5. J.N. Fayad, T. Baino, S.C. Parisier, Revision cochlear implant surgery: causes and outcome. Otolaryngol. Head Neck Surg. **131**, 429–432 (2004)

6. S.J. Rebscher, M. Heilmann, W. Bruszewski, N.H. Talbot, R.L. Snyder, M.M. Merzenich, Strategies to improve electrode positioning and safety in cochlear implants. IEEE Trans. Biomed. Eng. **46**, 340–352 (1999)

7. K.S. Min, A study on the liquid crystal polymer-based intracochlear electrode array, Thesis Seoul National University, 2014

8. B. Bhushan, Adhesion and stiction: mechanisms, measurement techniques, and methods for reduction. J. Vac. Sci. Technol., B **21**, 2262–2296 (2003)

9. C.J. Van Oss, R.J. Good, M.K. Chaudhury, The role of van der Waals forces and hydrogen bonds in "hydrophobic interactions" between biopolymers and low energy surfaces. J. Colloid Interface Sci. **111**, 378–390 (1986)

10. J.A. Bierer, S.M. Bierer, J.C. Middlebrooks, Partial tripolar cochlear implant stimulation: spread of excitation and forward masking in the inferior colliculus. Hear. Res. **270**, 134–142 (2010)

11. D. Vellinga, J.J. Briaire, D.M.P. van Meenen, J.H.M. Frijns, Comparison of multipole stimulus configurations with respect to loudness and spread of excitation. Ear Hear. **38**, 487–496 (2017)

12. D. Vellinga, S. Bruijn, J.J. Briaire, R.K. Kalkman, J.H.M. Frijns, Reducing interaction in simultaneous paired stimulation with CI. PLoS ONE **12**, e0171071 (2017)

13. J.A. Bierer, Threshold and channel interaction in cochlear implant users: evaluation of the tripolar electrode configuration. J. Acoust. Soc. Am. **121**, 1642–1653 (2007)

14. A. Kinloch, *Adhesion and Adhesives: Science and Technology* (Springer Science & Business Media, 2012)

15. J. Ordonez, M. Schuettler, C. Boehler, T. Boretius, T. Stieglitz, Thin films and microelectrode arrays for neuroprosthetics. MRS Bull. **37**, 590–598 (2012)

16. A.S. Widge, Self-assembled monolayers of polythiophene "Molecular wires": a new electrode technology for neuro-robotic interfaces, Thesis Carnegie Mellon University, 2007

17. W.-S. Kim, I.-H. Yun, J.-J. Lee, H.-T. Jung, Evaluation of mechanical interlock effect on adhesion strength of polymer–metal interfaces using micro-patterned surface topography. Int. J. Adhes. Adhes. **30**, 408–417 (2010)

18. L. Reclaru, J.M. Meyer, Study of corrosion between a titanium implant and dental alloys. J. Dent. **22**, 159–168 (1994)

19. A. Vanhoestenberghe, N. Donaldson, The limits of hermeticity test methods for micropackages. Artif. Organs **35**, 242–244 (2011)

20. K. Aihara, M.J. Chen, C. Cheng, A.V.H. Pham, Reliability of liquid crystal polymer air cavity packaging. IEEE Trans. Compon. Packag. Manuf. Technol. **2**, 224–230 (2012)

21. A.-V. Pham, Packaging with liquid crystal polymer. IEEE Microwave Mag. **5**, 83–91 (2011)

22. J. Jeong, S.H. Bae, J.-M. Seo, H. Chung, S.J. Kim, Long-term evaluation of a liquid crystal polymer (LCP)-based retinal prosthesis. J. Neural Eng. **13**, 025004 (2016)

23. M. Celina, K.T. Gillen, R.A. Assink, Accelerated aging and lifetime prediction: review of non-Arrhenius behaviour due to two competing processes. Polym. Degrad. Stab. **90**, 395–404 (2005)

24. D.W.L. Hukins, A. Mahomed, S.N. Kukureka, Accelerated aging for testing polymeric biomaterials and medical devices. Med. Eng. Phys. **30**, 1270–1274 (2008)

25. W. Chun, N. Chou, S. Cho, S. Yang, S. Kim, Evaluation of sub-micrometer parylene C films as an insulation layer using electrochemical impedance spectroscopy. Prog. Org. Coat. **77**, 537–547 (2014)

26. A.A.C.M. Lopez, S. Mitra, M. Welkenhuysen, W. Eberle, C. Bartic, R. Puers, R.F. Yazicioglu, G.G.E. Gielen, An implantable 455-active-electrode 52-channel CMOS neural probe. IEEE J. Solid-State Circuits **49**, 248–261 (2014)

27. S.W. Lee, K.S. Min, J. Jeong, J. Kim, S.J. Kim, Monolithic encapsulation of implantable neuroprosthetic devices using liquid crystal polymers. IEEE Trans. Biomed. Eng. **58** (2011)

28. T.M. Gwon, J.H. Kim, G.J. Choi, S.J. Kim, Mechanical interlocking to improve metal–polymer adhesion in polymer-based neural electrodes and its impact on device reliability. J. Mater. Sci. **51**, 6897–6912 (2016)

29. G.-T. Hwang, D. Im, S.E. Lee, J. Lee, M. Koo, S.Y. Park et al., In vivo silicon-based flexible radio frequency integrated circuits monolithically encapsulated with biocompatible liquid crystal polymers. ACS Nano **7**, 4545–4553 (2013)

30. V. Sundaram, V. Sukumaran, M.E. Cato, F. Liu, R. Tummala, J.D. Weiland et al., High density electrical interconnections in liquid crystal polymer (LCP) substrates for retinal and neural prosthesis applications, in *2011 IEEE 61st Electronic Components and Technology Conference (ECTC)* (2011), pp. 1308–1313
31. T.M. Gwon, C. Kim, S. Shin, J.H. Park, J.H. Kim, S.J. Kim, Liquid crystal polymer (LCP)-based neural prosthetic devices. Biomed. Eng. Lett. **6**, 148–163 (2016)
32. A. Leng, H. Streckel, K. Hofmann, M. Stratmann, The delamination of polymeric coatings from steel Part 3: Effect of the oxygen partial pressure on the delamination reaction and current distribution at the metal/polymer interface. Corros. Sci. **41**, 599–620 (1998)
33. A. Leng, H. Streckel, M. Stratmann, The delamination of polymeric coatings from steel. Part 2: First stage of delamination, effect of type and concentration of cations on delamination, chemical analysis of the interface. Corros. Sci. **41**, 579–597 (1998)
34. A. Leng, H. Streckel, M. Stratmann, The delamination of polymeric coatings from steel. Part 1: Calibration of the Kelvinprobe and basic delamination mechanism. Corros. Sci. **41**, 547–578 (1998)
35. S.W. Lee, J.M. Seo, S. Ha, E.T. Kim, H. Chung, S.J. Kim, Development of microelectrode arrays for artificial retinal implants using liquid crystal polymers. Invest. Ophthalmol. Vis. Sci. **50**, 5859–5866 (2009)
36. K. Aihara, M.J. Chen, A.V. Pham, Development of thin-film liquid-crystal-polymer surface-mount packages for *Ka*-band applications. IEEE Trans. Microw. Theory Tech. **56**, 2111–2117 (2008)

Chapter 5
Conclusion

In the present study, we developed an LCP-based cochlear electrode array for atraumatic deep insertion using a novel design scheme and an advanced LCP-based, microfabrication process. A tapered, multi-layer LCP substrate with 16-channel interconnection on an LCP film was successfully fabricated. Variable layers (depending on the part of the electrode array) were utilized to achieve flexibility at the tip and sufficient rigidity at the base. This design enables safer insertion with less insertion force. Dislocation of the tip of the electrode array into the scala vestibule was observed in one of five trials in human temporal bone insertion studies, whereas there was no observable trauma at the basal turn. These are more atraumatic results compared to the conventional, wire-based, uniformly thick LCP-based electrode. The initial functional EABR recording data were obtained via acute electrical stimulation in the guinea pig model. The data revealed that the hearing preservation and tissue reaction are comparable with conventional electrode arrays four weeks after in vivo implantation.

We also proposed an electrode site design for local tripolar stimulation to increase spectral resolution in multi-channel, polymer-based neural electrode arrays. Simulation results show the possibility of the proposed electrode site design and focused current stimulation. Multi-layered electrode sites, wherein center sites for stimulation are located on the bottom layer and auxiliary electrode sites are on the side wall of the upper layer of the LCP films, were realized using laser micromachining and via structures. The effect of the structure of electrode sites is evaluated using in vivo measurements of the spreading of current stimulation.

Long-term device reliability is a significant issue in polymer-based neural prostheses for clinical uses. The long-term reliability of LCP-based implantable devices was evaluated using analytical calculations, helium leak rate measurements, blister tests, and accelerated soak tests. Surface modifications, such as those by oxygen plasma and high-pressure lamination, have been used to improve the interlayer bond strength and were shown to enhance device reliability levels. In this dissertation, technical strategies to enhance reliability of the polymer-based device were suggested. An applicable fabrication method for polymer-based implantable neural interfaces relies

© Springer Nature Singapore Pte Ltd. 2018
T. M. Gwon, *A Polymer Cochlear Electrode Array: Atraumatic Deep Insertion, Tripolar Stimulation, and Long-Term Reliability*, Springer Theses,
https://doi.org/10.1007/978-981-13-0472-9_5

on mechanical interlocking at the interface between the polymer and a noble metal. Microscale undercut metal patterning using photolithography and electroplating are interlocked with LCPs without voids using a thermo-compression bonding process. Results of the peel test and cross-sectional images after device failures corroborate the effect of mechanical interlocking with the fact that the interlocked LCP-metal interface is intact while another interface is delaminated without interlocking, though it is difficult to estimate the lifetime accurately. Mechanical interlocking has the effect of preventing water ingress into the exposed electrode site window where the LCP-metal interface exists. The proposed method using mechanical interlocking at the electrode site not only helps LCP-based neural prosthetic devices gain widespread acceptance as reliable, long-term implantable devices, but also shows they are applicable to other polymer-based devices. Additional strategies to use dielectric materials with LCPs were proposed and their feasibility in the LCP fabrication process was investigated in preliminary tests. LCPs and dielectric materials surface-treated with oxygen plasma were bonded using a thermo-compression lamination process, resulting in reliable adhesion strength. Metal migration owing to heat and pressure in the lamination process can be prevented using a hybrid structure of LCPs and dielectric materials.

Technical results, performance tests, and long-term reliability data indicate a promising future for safe, effective, long-term, reliable LCP-based neural electrode arrays, despite several remaining challenges and clinical use verifications.

Printed by Printforce, the Netherlands